T0177765

The Foundations of Modality

The Foundations of Modality

From Propositions to Possible Worlds

PETER FRITZ

OXFORD
UNIVERSITY PRESS

OXFORD
UNIVERSITY PRESS

Great Clarendon Street, Oxford, OX2 6DP,
United Kingdom

Oxford University Press is a department of the University of Oxford.
It furthers the University's objective of excellence in research, scholarship,
and education by publishing worldwide. Oxford is a registered trade mark of
Oxford University Press in the UK and in certain other countries

© Peter Fritz 2023

Published in the United States of America by Oxford University Press
198 Madison Avenue, New York, NY 10016, United States of America

British Library Cataloguing in Publication Data

Data available

Library of Congress Control Number: 2023937241

ISBN 978–0–19–287002–5

DOI: 10.1093/oso/9780192870025.001.0001

Printed and bound by
CPI Group (UK) Ltd, Croydon, CR0 4YY

Links to third party websites are provided by Oxford in good faith and
for information only. Oxford disclaims any responsibility for the materials
contained in any third party website referenced in this work.

Contents

III. NECESSITY

IV. WORLDS

V. CONCLUSION

Acknowledgements

What separates the necessary from the merely contingently true? What separates the impossible from the merely contingently false? And what, if anything, are possible worlds? In this book, I try to answer these questions by articulating and motivating a foundational theory of modality in philosophy. The theory I end up with is not based on any one single big idea. Rather, many ideas—some bigger, some smaller—work together to make up a story about how we might understand modality in philosophy. Some of these ideas are my own, and new, but most of them are due to others, and have been described in print before. The aim of this book is therefore only partly to present new ideas. To at least the same extent, the aim of this book is to present existing ideas, and to do so in a way which is accessible and which shows how they fit together to make up a coherent story about modality.

As a consequence, I owe large intellectual debts to a number of people for their work on which this book draws. The bibliography tells of the publications without which I could not have arrived at the theory I present here. I also owe large intellectual debts to a number of people for influences on this book which are not recorded in the bibliography. I am grateful to all of them, and I will try to mention as many of them as I can here. For those I have forgotten, I add my apologies to my thanks.

Timothy Williamson and Cian Dorr supervised my dissertation, which I had intended to write on the topic of this book. It was in no small part their thoughtful criticisms which showed me the limitations of my initial thoughts in this direction, and which led me to abandon them. Instead, I ended up working on modal metaphysics, taking the notion of metaphysical necessity which I had intended to question mostly as given. As it turned out, it was the logical and philosophical methods I learned while taking this detour that eventually allowed me to tackle my initial questions in a more fruitful way. I am therefore deeply grateful to them for their guidance, and for everything else I have learned from them. Their influence permeates this book, and I could not have written it without their teaching.

I could also not have written this book without having participated in an online reading group with Andrew Bacon, Michael Caie, Cian Dorr, Jeremy Goodman, and Harvey Lederman over the last few years. I am deeply grateful for everything I have learned from them about higher-order logic and propositional granularity through discussing their work in progress and other papers and books. I am equally grateful for being able to discuss with them drafts of the material in this book at multiple stages. Without their comments and questions, the end result would have been much poorer.

Johannes Marti, Alexander Roberts, Sam Roberts, Juhani Yli-Vakkuri, and two readers for Oxford University Press also read drafts of material in this book, and sent me very helpful written comments, which led to substantial improvements of the book, and for which I am very grateful.

I presented material from this book at a number of conferences and seminars, at the Australian Catholic University, Humboldt University Berlin, University of Birmingham, King's College London, University of Melbourne, University of Oslo, Oxford University, Princeton University, University of Southern California, University of Toronto, and the University of Waikato. Drafts of this book also served as the basis of a graduate class at the University of Oslo in 2021 and a number of meetings of a reading group on higher-order logic at the Australian Catholic University. I would like to thank the audiences and participants on all of these occasions for their comments, questions, and discussion, in particular Andrew Bacon, Sam Baron, Nathaniel Baron-Schmitt, John Bigelow, Kyle Blumberg, David Builes, Tim Button, Michael Caie, Sam Carter, Andreas Ditter, Antony Eagle, Stephen Finlay, Vera Flocke, Rachel Fraser, Dagfinn Føllesdal, Dmitri Gallow, Simon Goldstein, Jeremy Goodman, Zachary Goodsell, Volker Halbach, John Hawthorne, Benj Hellie, Wesley Holliday, Amirhossein Kiani, Philip Kremer, Tamar Lando, Martin Leckey, Øystein Linnebo, Jon Litland, Annina Loets, Ofra Magidor, Matthew Mandelkern, Dan Marshall, Beau Madison Mount, Sarah Moss, Christopher Masterman, Alexander Paseau, Martin Pickup, Agustín Rayo, Greg Restall, Michael Rieppel, David Ripley, Alexander Roberts, Sam Roberts, Gideon Rosen, Sven Rosenkranz, Gillian Russell, Bernhard Salow, Stewart Shapiro, Lukas Skiba, James Studd, Kai Tanter,

Barbara Vetter, Lisa Vogt, Richard Woodward, Juhani Yli-Vakkuri, and Jin Zeng.

While writing this book, I benefited from the congenial work environment at the Dianoia Institute of Philosophy at ACU, for which I am grateful to all of my colleagues at the institute. I am especially grateful to Gillian Russell for organizing weekly writing sessions which became invaluable to me as the book was nearing completion.

I would also like to thank my editor, Peter Momtchiloff, for his support and advice, and more generally everyone involved at Oxford University Press for their care in producing this book.

Introduction

The Ubiquity of Modalities

In ordinary life, many of our concerns pertain to *what* things there are and *how* those things are. For example, if you want to go to the market to buy barley, you will be concerned with whether there is barley available at the market, and whether such barley is of good quality. Many of the concerns of the sciences pertain to questions which are similar in kind, although more general. For example, astrophysics addresses whether there are pulsars, and whether such pulsars are neutron stars. Many of the concerns of metaphysics are similar as well, but take another step in the direction of generality. For example, metaphysics addresses whether there are properties, and whether such properties are sets of possibilia (as proposed by Lewis 1986).

However, we do not only care about what there is and how these things are. For example, when going to the market to buy barley, you might also be concerned with questions such as the following:

- Did I buy barley *last week*?
- *Should* I buy barley?
- Am I *able* to walk to the market before it closes?
- Does Kushim *want* me to buy barley?
- Did Kushim *say* that barley is in demand?
- Did Kushim *tell me to* buy barley?
- Does Kushim *know* I am going to the market?
- Does Kushim *believe* I am at home?
- Am I buying barley *because* I am hungry?
- *Would* I have gotten better barley *if* I had gone to the market earlier?

It is natural to think of these further questions as going beyond what there is and how these things are. For example, the question whether I

The Foundations of Modality: From Propositions to Possible Worlds. Peter Fritz, Oxford University Press.

am buying barley because I am hungry goes beyond asking about how I in fact relate to barley and how I am feeling: beyond asking whether me buying barley and me being hungry are the case or not, the question pertains to a possible explanatory link between me being hungry and me buying barley. This is a general feature of these further questions: they all involve constructions which can be construed as modifying sentential expressions, while being sensitive to more than truth-value. In logical terms, these constructions can therefore be understood as non-truth-functional operators. The study of such operators is the purview of modal logic. In a wide sense of the word, all of these questions can therefore be understood as *modal*, and as involving some (non-truth-functional) *modality*. For example, the question whether I should buy barley involves a *deontic modality*, expressed by the word "should"; the question whether Kushim knows that I am going to the market involves an *epistemic modality*, expressed by "Kushim knows"; the question whether I am buying barley because I am hungry involves a (binary) *explanatory modality*, expressed by "because"; and so on.

Such modal concerns are pervasive in our everyday lives, as the above examples illustrate. Every day, virtually every person who has any thoughts at all entertains a multitude of modal thoughts. In fact, many of our everyday thoughts involve not just one modality, but several iterated modalities. Already some of the examples above involved a temporal modality—of having been the case in the past— in addition to a second modality of saying, telling, knowing, or believing. Indeed, we regularly entertain questions which from a logical perspective involve a staggering amount of modal complexity. For example, on the way to the market, you might easily recall an earlier conversation, and wonder:

Did Kushim say that I shouldn't always buy barley just because I think that we might be out of barley soon?

This involves several temporal modalities (expressed by past tense and words like "always" and "soon"), a modality of saying ("say"), a deontic modality ("should"), an explanatory modality ("just because"), a modality of thinking ("think"), and an epistemic modality ("might").

As one might expect, these ubiquitous modal concerns also arise in the sciences and in metaphysics. Astrophysics, for example, addresses questions about the distant past of the universe through observations about cosmic microwave background radiation. This science is therefore substantially concerned with what *was* the case, in addition to what *is* the case. For other examples, note that it was an important astrophysical insight that the expansion of the universe *explains* the redshift of galaxies, and similarly that rotating universes are *possible* according to the theory of general relativity. As in the cases of ordinary life and scientific inquiry, many examples of metaphysical questions and claims can be found which involve various forms of modality. For example, the debate between necessitists—who think that it is a necessary matter what there is—and contingentists—who deny this—essentially involves the modal notion of *necessity*.

Ideological Disputes

In metaphysics, modalities have also become the focus of attention themselves. This is partly because modalities play an important role in philosophical theorizing more generally. One might say that they provide the ideology to articulate the very ambitions of philosophy.

By way of example, consider the mind-body problem. This can be articulated roughly as the question of how the mind relates to the body. For example, how does being in pain relate to various states of nervous systems? One might begin answering this question by investigating co-occurrences of states of pain with various states of nervous systems. But one might naturally go further, and ask whether someone being in a certain kind of pain is *necessitated* by their nervous system being in the relevant state; whether the former is *grounded* in the latter; and whether the former is *identical* to the latter.

In many cases, the choice of modalities employed in philosophical investigations substantially influences the shape of the resulting theories. The range of cases goes beyond metaphysics, as the example of the mind-body problem just given already illustrates, being as much a question of the philosophy of mind as of metaphysics. Another example is the philosophy of science, as argued recently in detail by Sider (2020). Partly as a consequence of this, a substantial amount of metaphysics (and metametaphysics, in the sense of Chalmers et al. 2009) concerns modalities directly. In many cases, this has the form of investigating a particular modality and its properties. The earlier example of the dispute between necessitists and contingentists, discussed in detail by Williamson (2013), illustrates this for necessity.

In other cases, philosophers step back from the features of any particular modality, and ask more generally which modalities should be employed in metaphysics, and philosophy more generally. For example, appeals to necessity were famously criticized by Quine (1980 [1953]), and defended by Kripke (1980 [1972]). More recently, Fine (2001) has argued for the introduction of a new explanatory relation of *ground(ing)* into metaphysical discourse. Given the generality of the ambit of metaphysics, one might wonder how anyone could argue that a given modality should be *excluded* from metaphysical discussion. Typically, the argument is not that the purview of metaphysics is too limited to include the relevant modality, but that the modality is in some sense not in good standing. This argument can take a number of forms. The following two forms are especially important.

First, modal terms expressing the relevant notions may be argued to be problematically context-sensitive, vague, or ambiguous. For example, the claim that Kushim *may* buy barley can be used in very different ways: It may be used to express a state of ignorance; i.e., to say that according to a certain contextually determined state of knowledge, it is not ruled out that Kushim buys barley. Or it may be used to express permission, to say that certain contextually determined prescriptions do not prohibit Kushim from buying barley. Similar examples can be given for many of the other modal expressions mentioned above.

For the purposes of metaphysics, such multiplicities of meaning are problematic. If a modality is meant to play an important role in our metaphysical theorizing, then we need a stable way of expressing it. If ordinary modal terms are associated with multiple meanings, they cannot on their own serve to pick out the desired modal notions. Of course, one might attempt to sharpen the relevant terms. For example, in metaphysics, it is common to talk of "metaphysical modality" in order to isolate the intended notion among the various candidates. But typically, the way these technical terms are introduced does not obviously guarantee that a single modality is picked out. It is therefore often controversial that these attempts at sharpening the common expressions succeed.

Second, many modal terms concern the representation of reality by various cognitive agents. For example, the claim that Kushim *knows* that you are going to the market concerns Kushim's epistemic state. The relevant epistemic state essentially involves Kushim's cognitive representation of you going to the market. However, metaphysics is concerned with a (very general) account of *reality*, rather than our *representation of reality*. To employ epistemic modalities in metaphysical theorizing therefore carries the risk of confusing reality with the representation of reality. Admittedly, at this level of abstraction, this is a relatively vague worry, since any representation of reality is part of reality itself.

To sharpen the concern, it is helpful to return from modalities to more ordinary ascriptions of how things are. By way of example, consider the claim that Superman is tall. This is a straightforward example of a description of reality (at least in the fictional universe of Superman), ascribing a property to an object. Since Superman is Clark Kent, it follows that Clark Kent is tall. In contrast, consider the claim that Superman is feared (by some contextually salient antagonist). It is much less clear that it follows that Clark Kent is feared as well. Whatever the full semantic story is, there are contexts in which it is appropriate to state that Superman is feared while it is not appropriate to state that Clark Kent is feared. But it does not follow that there is a property—being feared by the relevant agent— which Superman has but which Clark Kent lacks. After all, Superman is Clark Kent, so any property Superman has, Clark Kent has as well. Whatever the relevant difference is, it must be put down to a difference in two *representations* of Superman, rather than to differences between two objects.

Something similar happens with modalities. Since Superman is Clark Kent, for Superman to fly is for Clark Kent to fly. But someone might know that Superman flies without knowing that Clark Kent flies. To say that someone knows that Superman flies therefore need not only concern Superman flying, but might also concern that person's representation of Superman flying. In this sense, to say that Kushim knows that you are going to the market need not just attribute to you going to the market some modality of being known, just like saying that Superman is feared need not just attribute to Superman some property of being feared.

In many cases in which an agent is explicitly mentioned, such as ascriptions of knowledge, belief, saying, and telling, it is relatively easy to see that a modal construction might involve sensitivity to representation. In other cases, this is harder to tell. As mentioned above, there are epistemic uses of words like "may," "might," and "must," which attribute some epistemic state to a contextually salient agent. Such epistemic modals do not wear their potential relativity to an agent's representational perspective on their sleeve. This raises the worry that modal terms employed in metaphysics might display a problematic and unintended sensitivity to representational features. As an example, consider the case of grounding in metaphysics. Similar to the case of metaphysical necessity, defendants of grounding concede that words like "because" can be used in multiple ways, and make various stipulations in order to sharpen its meaning to arrive at a unique notion of metaphysical grounding. But even assuming that this succeeds, if every way of using "because" is sensitive to the representational states of an implicit agent, then no modality has been identified which can play the intended role in metaphysics.

As a consequence of these two and other concerns about various modal notions, there is substantial disagreement about which such notions should be employed in metaphysics, and philosophy more generally. As the two concerns discussed above illustrate, a considerable portion of these concerns is semantic or metasemantic, and pertains to the question whether various ways of sharpening ordinary terms succeed in pinning down unique modalities which are not representationally sensitive. Such questions are notoriously difficult to settle. As a consequence, many of the debates result in unsatisfying stalemates, with one party simply asserting that they *understand* a term like "metaphysical necessity" or "grounding," and the opposing party denying equally firmly that they understand it (with the implication that the first party is under an illusion of understanding).

Identity-First Metaphysics

This book tries to make progress on these fundamental metaphysical questions. It does so by shifting attention away from trying to adjudicate between conflicting intuitions about the intelligibility of various proposed modal notions. Instead, the approach will be to develop a more basic understanding of modalities in metaphysics, and then to draw consequences from this to particular modalities. Developing this more basic understanding involves two crucial components. They are inspired by recent work in metaphysics which uses higher-order logic, such as Williamson (2013), and which focuses on identity, such as Dorr (2016).

First, we will adopt an artificial language which allows us to formally regiment talk of modalities. This will fall out of a more general program of developing such a language which allows us to formally regiment talk of entities like propositions, properties, and relations. Modalities have most successfully been studied using the tools of modal logic, and the language which we will use below is an extension of modal logic which includes the quantificational resources needed to make philosophical progress. The language includes variables which take the position of modal operators, and quantifiers binding them. Since modal operators are sentential operators, it also includes variables which take the position of their arguments, i.e., formulas, and quantifiers binding them. In English, these quantifiers binding variables in sentence position are naturally described as quantifying over propositions. Quantifiers binding variables in operator position can correspondingly be described as quantifying over properties of propositions. According to this way of rendering the formal language in English, a modality like necessity is a property of propositions: to say that it is necessary that 2+2=4 is to attribute the property of being necessary to the proposition that 2+2=4.

Second, we will make progress on understanding the space of modalities by developing a theory of individuation of propositions. The idea behind this approach is the following. Even those who disagree about which modalities are in good standing should agree that there is an unproblematic notion of propositions being identical; after all, every proposition is identical to itself and to no others. Agreeing *that there is* an unproblematic notion of propositions being identical does not require agreeing on *what it takes* for propositions to be identical. To the contrary, different viewpoints in metaphysics employing different distinctive vocabularies often come with corresponding—and correspondingly incompatible—commitments on the individuation of propositions. For example, grounding is often taken to lead to a fine-grained individuation of propositions which is denied by those who focus on necessity. This is one way in which insights into propositional individuation can be harnessed to yield insights into what modalities there are: there cannot be any modality which distinguishes identical propositions. Therefore, disagreements about propositional identity are one crucial fact which the shift to propositional individuation aims to exploit. Propositional identity provides a neutral arena in which different sides can discuss substantial metaphysical questions without becoming bogged down in semantic and metasemantic disputes about the status of debated ideology. In this sense, this book exemplifies what one might call *identity-first metaphysics.*

With these two strategies, this book will argue for a particular theory of modalities in metaphysics. The resulting view is quite orthodox, in that a distinguished modality of metaphysical necessity plays a crucial role. Propositions are individuated modally, with propositions being identical just in case they are metaphysically necessarily equivalent. Metaphysical necessity can also be understood in terms of being true in all possible worlds, with possible worlds being accounted for in purely propositional terms. Although the view is familiar, the argument toward it involves many much less well known steps.

Outline

The parts and chapters of this book correspond to the steps in the argument for the view of metaphysics to be defended. Although it is possible to read some of the chapters in isolation, the argument, like the book, has a strictly linear structure. Part I introduces the formal language; Part II develops a theory of propositional individuation; Part III isolates and investigates the central modality of necessity; Part IV develops a theory of possible worlds; and Part V concludes. Each part has two chapters.

Part I: Language. Chapter 1 discusses how talk of propositions, properties, and relations should be regimented in metaphysics, and argues for a higher-order logical language. It further proposes a way of introducing this language in such a way that its expressions have intended interpretations. The resulting language can therefore not just be used as a formal tool, but as a meaningful language in which metaphysical debates can be carried out. In this language, modalities can be understood as properties of propositions. Chapter 2 introduces more rigorously a fragment of the language. In order to keep the formalities in this book to a minimum, this more limited language only includes the features which are essential for talking about propositions and their properties (i.e., modalities).

Part II: Identity. Propositional identity can be understood as a binary property of propositions. With the formal language at hand, the question of propositional identity can therefore be investigated systematically. Chapter 3 starts with a simple argument for a very fine-grained, structured individuation of propositions. It is shown that this immediately runs into serious problems, as the natural formal regimentation turns out to be formally inconsistent. These difficulties extend to natural theories of grounding, which are similarly shown to be inconsistent. This provides the first example of how focusing on identity can lead to progress on questions of metaphysical ideology. The inconsistency of fine-grained theories of propositional individuation motivates investigating coarse-grained alternatives. Chapter 4 turns to such theories, and uses general criteria of theory choice to motivate one specific such theory. According to this theory, propositions are identical if they are expressed by sentences which can be shown to be materially equivalent using certain principles of a broadly logical character.

Part III: Necessity. Chapter 5 argues on the basis of this theory of propositional individuation that metaphysical necessity can be identified in terms of identity. According to the theory, there is a distinguished tautological proposition. The modality of being identical to this tautology can then be shown to be distinguished as the broadest necessity. With an argument for the claim that metaphysical necessity is the broadest necessity, it follows that a proposition is metaphysically necessary just in case it is the tautological proposition. A number of substantial metaphysical commitments follow from this identification of metaphysical necessity. For example, it follows that the propositional logic of metaphysical necessity includes the well-known modal logic S4, and that identical propositions are necessarily identical. However, it does not follow that the propositional logic of metaphysical necessity includes the strong modal logic S5, which adds to S4 the claim that what is possible is necessarily possible. Similarly, it cannot be shown that distinct propositions are necessarily distinct. Chapter 6 shows that these additional claims can be derived once an actuality operator is added to the language.

Part IV: Worlds.   One consequence of the proposed identification of necessity is that propositions which are necessarily equivalent are identical. This is a claim familiar from possible worlds models of propositions, in which propositions are identified with sets of possible worlds, and necessity amounts to truth in all possible worlds. Chapter 7 considers whether such formulations in terms of possible worlds are also available on the present view. This chapter develops an abstract theory of possible worlds, and concludes that standard theorizing in terms of possible worlds stands and falls with a principle about propositions which can be stated just in terms of necessity. This is the principle that, necessarily, there is a true proposition which is maximal in the sense of strictly implying, for every proposition, either it or its negation. Chapter 8 shows that this principle cannot be derived from the theory developed up to this point. However, this chapter also fills the remaining gap by showing that the principle can be derived once a plural form of propositional quantification is added to the language under consideration.

Part V: Conclusion.   With the notion of possible worlds put on a solid foundation, the resulting view is an instance of a very well-known modal approach to metaphysics. On this view, a modality of metaphysical necessity occupies a central role, and satisfies a very strong and simple modal logic. Chapter 9 sums up the resulting picture, and uses its coarse-grained individuation of propositions to address a widespread worry about the use of higher-order logic in metaphysics, concerning the inability to express generality across different types of variables. Chapter 10 concludes by looking ahead at important questions not discussed in this book. It will be argued that the main challenge lies in developing convincing alternative proposals to the relatively familiar picture endorsed here.

PART I
LANGUAGE

1

Breaking the Domination of the Word over the Human Spirit

1.1 Introduction

Propositions and modalities can be thought of as being of a kind with universals like properties and relations. To understand propositions and modalities, it will therefore be useful to start by reflecting more generally on how to think about properties and relations, alongside propositions and modalities. This is the topic of the present chapter. The next chapter applies the conclusions of this chapter to propositions and modalities.

By way of illustration, consider the case of properties. Naive reasoning about properties quickly leads to inconsistency. For example, take the property of being a property which doesn't have itself as one of its properties. Does it have itself as one of its properties? If it does, it doesn't, and if it doesn't, it does—a contradiction. This is simply the property-theoretic analog of the observation, made by Russell (1903), that naive set theory is inconsistent. There is therefore a clear need to improve upon naive reasoning about properties.

As elsewhere in philosophy when naive reasoning leads to inconsistency, it is useful to regiment the discussion in terms of a formal language. Doing so helps overcome various potential semantic and metasemantic defects of natural language. A simple case of such a defect is structural ambiguity, which is easily eliminated in formal languages. For example, first-order logic disambiguates "Everyone loves someone" using $\forall x \exists y L x y$ and $\exists y \forall x L x y$. More substantially, natural language expressions used to talk about properties and similar kinds of entities may be ambiguous, similar to the case of "bank," which can be used to talk about either financial institutions or rivers. A formal language whose interpretation is constrained in more explicit ways may resolve such ambiguities.

The Foundations of Modality: From Propositions to Possible Worlds. Peter Fritz, Oxford University Press.
© Peter Fritz 2023. DOI: 10.1093/oso/9780192870025.003.0002

The practice of improving our ideological resources as part of philosophy has recently become the object of attention of philosophers such as Cappelen (2018), who calls it "conceptual engineering." However, as Cappelen notes, the practice has been part of philosophy since its earliest days. This applies as well to the use of formal languages, which first arose in the development of symbolic logic. Frege (1879) expresses this clearly in his *Begriffsschrift*, which may be considered the founding document of modern logic. In its preface, he writes of the formal language developed there:

> If it is one of the tasks of philosophy to break the domination of the word over the human spirit by laying bare the misconceptions that through the use of language often almost unavoidably arise concerning the relations between concepts and by freeing thought from that with which only the means of expression of ordinary language, constituted as they are, saddle it, then my ideography, further developed for these purposes, can become a useful tool for the philosopher.
>
> (van Heijenoort 1967, p. 7)

Section 1.2 develops some desiderata of a formal language of propositions, properties, relations, and modalities. Section 1.3 shows that the desiderata are inconsistent, along the lines of the inconsistency sketched informally above, and argues that one of the desiderata is less central than the others. This motivates introducing a formal language which satisfies the remaining desiderata. As a preliminary, section 1.4 discusses how formal languages can be introduced in a way which renders them meaningful. Section 1.5 introduces a language along these lines, together with general principles of a logical character which partially serve to constrain the meanings of the formal language. The resulting language and principles constitute a well-known form of higher-order logic which is sometimes called "relational type theory."

For concreteness and brevity, we focus on properties from now on, assuming a general conception on which properties can have any finite arity. In the language we will eventually adopt, such properties arguably cover all cases of propositions, properties, relations, and modalities mentioned so far: propositions can be understood as nullary properties, *n*-ary relations can be understood as *n*-ary properties, for any $n \geq 2$, and modalities can be understood as properties of propositions.

1.2 Desiderata of Property Talk

One of the most commonly used formal language in philosophy is the language of first-order logic. This language provides ways of making simple claims involving properties, such as the claim that Jupiter is a planet, which attributes to Jupiter the property of being a planet. It is a clear desideratum of a formal language of properties to provide these resources, including a predicate of identity, and the ability to embed any statements under Boolean connectives and the ability to bind any variables by quantifiers. These logical connectives should obey the standard truth-conditions, so that in particular no truth-functional contradiction, such as any formula of the form $\varphi \leftrightarrow \neg\varphi$, is true. This motivates the first four desiderata:

BASIC PREDICATIONS: First-order predications and identity statements are formulas.

TRUTH-FUNCTIONS: Formulas are closed under Boolean connectives.

QUANTIFIERS: For any variable *x*, formulas are closed under universal and existential quantifiers binding *x*.

NON-CONTRADICTION: No truth-functional contradiction is true.

Here, a first-order predication may be assumed to consist of an *n*-ary predicate constant and a sequence of *n* individual terms. Individual terms may be taken to be either individual variables or individual constants. This leaves out function terms, which are often included in formulations of first-order logic, and which allow the formation of complex individual terms. The reason for leaving out function terms is just simplicity: function terms are not required to capture talk of properties.

NON-CONTRADICTION differs from the other three desiderata in that it is not a principle about the structure of the formal language under consideration, but about the truth of its formulas. Here, truth is not intended to be a notion which is relativized to any models, set-theoretic or otherwise. The present concern is to construct a formal language which *improves upon* natural languages. Thus, the formal language should serve the same purposes of natural languages, including the ability to assert its sentences and to express truths (and falsehoods) *simpliciter*. Naturally, the truth of a sentence in the formal language (i.e., a formula without free variables) will be relative to an interpretation of its non-logical constants. But such an interpretation may be specified informally, in a given context of use, for example by stating that an individual constant j refers to Jupiter. Furthermore, some formulas have neither free variables nor non-logical constants, such as $\forall x(x = x)$. Such sentences should simply be true or false.

Since Lewis (1983), it has become common to distinguish between two conceptions of properties, a *sparse* conception and an *abundant* conception. Basic physical properties, like the property of being an electron, are often regarded as falling under the sparse (as well as the abundant) conception of properties. In contrast, so-called "gerrymandered" properties, such as the property of being an electron or a moon of Jupiter, are usually considered as falling under the abundant but not the sparse conception of properties. In the time since Lewis's influential paper, it has become common to use talk of properties in the abundant sense, and to capture the sparse conception of properties by distinguishing certain properties as *natural*; see, e.g., Dorr and Hawthorne (2014). This will be the approach taken in the following: properties will be taken to be abundant, and the question whether there is a role to play for some of them to be distinguished as natural will be set aside for another occasion.

The usual examples of "gerrymandered" properties make clear that on the present abundant conception of properties, it is usually assumed that any statement may be used to identify a property, namely the property it attributes to t, for any given singular term t. For example, one may use the statement "Io is a moon of Jupiter" to identify a property which may be expressed by "being a moon of Jupiter." Furthermore, it is usually assumed that any term for a property extracted from a statement in this way may be used to attribute the property it expresses to something, for example by attributing to Io the property of being a moon of Jupiter. Finally, in such an attribution, no material distinction is usually made between attributing the relevant property to the relevant entity and the original statement. So, it is usually assumed that Io has the property of being a moon of Jupiter if and only if Io is a moon of Jupiter.

This motivates a second set of desiderata for property talk. Similar to the formalization of quantificational statements, the extraction of property terms is most naturally effected using variables. For example, in a basic predication Mji, the occurrence of i may be indicated using a variable x, leading to the formula Mjx. It is desirable to have a corresponding property term, which may be indicated by $[x.Mjx]$. To attribute the property expressed by $[x.Mjx]$ to the individual denoted by i, a corresponding syntactic construction is desirable, which may be indicated by $[x.Mjx]@i$. Finally, it is desirable for the attribution $[x.Mjx]@i$ to be materially equivalent to Mji. The following three items generalize these desiderata to arbitrary formulas and variables, and to n-ary relations for any $n \geq 0$.

EXTRACTION: For any formula φ and pairwise distinct variables $x_1, \ldots, x_n$, there is a corresponding property term $[x_1 \ldots x_n.\varphi]$.

ATTRIBUTION: If terms $t_1, \ldots, t_n$ can be substituted for pairwise distinct variables $x_1, \ldots, x_n$ in a formula φ, then there is a corresponding formula $[x_1 \ldots x_n.\varphi]@t_1 \ldots t_n$.

TRANSPARENCY: Formulas of the following form are true:

$$[x_1 \ldots x_n.\varphi]@t_1 \ldots t_n \leftrightarrow \varphi[t_1, \ldots, t_n/x_1, \ldots, x_n]$$

In TRANSPARENCY, $\varphi[t_1, \ldots, t_n/x_1, \ldots, x_n]$ is the result of simultaneously replacing every free occurrence of x_i in φ by t_i, for $i \leq n$. Here, it is assumed that t_i is free for x_i in φ; i.e., that no free occurrence of a variable in t_i becomes bound after the replacement. Note that in this desideratum, square brackets are used for two unrelated purposes, on the left hand side indicating property terms and on the right hand side specifying replacements.

A couple of clarifications concerning these desiderata are in order. First, one might wonder why property terms are not written using the common λ-notation, which we will employ from the next chapter on. The reason is two-fold: First, λ-terms are often used as complex predicates, and we will do so as well below. However, at this point we want to make no assumptions about the syntactic category to which property terms belong. Second, a principle like EXTRACTION is not meant to describe the specific syntax of any particular language, but instead to require a language to provide certain general features. That is, square brackets and the @ symbol are used here only to indicate terms playing a certain function. Languages satisfying these desiderata don't have to use these particular symbols in the relevant constructions. For example, the particular languages defined below use λ-terms to play the role which is described here using square brackets and the @ symbol.

Second, the three desiderata just stated can be instantiated using any formula φ, and using n pairwise distinct variables, for any natural number n including 0. This generality is motivated by examples similar to the ones mentioned above. For example, we might want to talk about the relation of being a *submoon*, where y is a submoon of x just in case y is a moon of some moon of x. Consequently, we might formulate the complex binary property term $[xy.\exists z(Mzy) \wedge Mxz]$. Using a nullary instance of EXTRACTION, we can also form the nullary property term $[.Mji]$. While talk of nullary properties may initially seem odd, these terms naturally serve to formalize talk of propositions, as we will see in more detail below.

General theorizing about properties requires the ability to make statements of generality about properties. From Io and Europa being moons of Jupiter, one naturally infers that there is some property which Io and Europa share, namely being a moon of Jupiter. In the first-order basis postulated by the first three principles, generality is captured by variable-binding quantifiers. It is therefore desirable to be able to replace any property term by a variable χ (which may or may not be an individual variable). Thus, from the well-formedness of

$$[x.Mjx]@i \wedge [x.Mjx]@e$$

one may infer the existence of a formula $\chi@i \wedge \chi@e$. By QUANTIFIERS, the variable χ may be bound by an (existential) quantifier:

$$\exists\chi(\chi@i \wedge \chi@e)$$

This motivates the following desideratum:

GENERALITY: Every property term can be replaced by some variable.

Note that EXTRACTION and GENERALITY together entail that there are terms which express properties of properties. For example, using the formula $\chi@i \wedge \chi@e$ provided by GENERALITY, one can infer with EXTRACTION that there is a property term $[\chi.\chi@i \wedge \chi@e]$. This is welcome, as it is desirable to have a way of talking about the property of being a property which is had by both Io and Europa. Furthermore, ATTRIBUTION allows this property term to be combined with $[x.Mjx]$ using @, with which TRANSPARENCY entails the truth of the following formula:

$$[\chi.\chi@i \wedge \chi@e]@[x.Mjx] \leftrightarrow [x.Mjx]@i \wedge [x.Mjx]@e$$

This is also welcome, as one naturally considers the property of being a moon of Jupiter to have the property of being a property which applies to both Io and Europa just in case the property of being a moon of Jupiter applies to both Io and Europa.

Given the ultimate aim of formalizing talk of propositions and modalities, it is also worth noting that the desiderata mentioned so far together entail that there are terms which express properties of nullary properties. Such terms can be used to capture talk of modalities. For example, we have noted above that by EXTRACTION, there is a nullary property term $[.Mji]$. From this, it follows with ATTRIBUTION that there is a formula $[.Mji]@$; this is naturally read as saying that the proposition that Io is a moon of Jupiter is true. With GENERALITY, the property term can be replaced by a variable χ, so there is a formula $\chi@$. Using EXTRACTION again, there is a property term $[\chi.\chi@]$; we can think of this as standing for the property of being a true proposition. On a very wide conception of modalities, on which even Boolean connectives express (edge cases of) modalities, we can think of this as the modality of truth.

Finally, it is desirable to be able to attribute a property to itself. By way of example, consider the property of self-identity. It is natural to think that everything is self-identical, including the property of being self-identical. Thus, the property of being self-identical should have the property of being self-identical. By BASIC PREDICATIONS, TRUTH-FUNCTIONS, and EXTRACTION, there is a property term $[x.x = x]$. It is therefore desirable to attribute $[x.x = x]$ to itself, using a formula of the form $[x.x = x]@[x.x = x]$.

SELF-APPLICATION: For every property term π, $\pi@\pi$ is a formula.

This may naturally be generalized to allow $\pi@\rho$ to be a formula for any property terms π and ρ. For present purposes, it makes no difference whether SELF-APPLICATION or this more general desideratum is used.

The motivations given here for the various desiderata are not conclusive. They are not *meant* to be conclusive: the aim of this section was only to use natural ways of naive reasoning about properties to motivate features which a formal language for reasoning about properties would ideally provide. However, as we will now see, they cannot all be satisfied, since they are inconsistent.

1.3 Inconsistency

The inconsistency of the desiderata follows along the lines of Russell's observation that naive set theory is inconsistent, already mentioned at the beginning of this chapter:

By BASIC PREDICATIONS, $x = x$ is a formula. By EXTRACTION, it follows that $[x.x = x]$ is a property term. Thus SELF-APPLICATION guarantees the existence of the following formula:

$$[x.x = x]@[x.x = x]$$

With GENERALITY, it follows that there is a variable χ for which $\chi@\chi$ is a formula. With TRUTH-FUNCTIONS, $\neg\chi@\chi$ is a formula, and so with EXTRACTION, $[\chi.\neg\chi@\chi]$ is a property term. Applying SELF-APPLICATION to this term produces the following formula:

$$[\chi.\neg\chi@\chi]@[\chi.\neg\chi@\chi]$$

With TRUTH-FUNCTIONS, this can be embedded under Boolean connectives:

$$[\chi.\neg\chi@\chi]@[\chi.\neg\chi@\chi] \leftrightarrow \neg[\chi.\neg\chi@\chi]@[\chi.\neg\chi@\chi]$$

Finally, TRANSPARENCY guarantees that this true. But since it is an instance of the truth-functionally inconsistent schema $\varphi \leftrightarrow \neg\varphi$, this contradicts NON-CONTRADICTION.

One must therefore drop at least one of the desiderata. ATTRIBU-TION and QUANTIFIERS are not involved in the argument. BASIC PRED-ICATIONS, TRUTH-FUNCTIONS, and NON-CONTRADICTION are essential components of first-order logic, which has been an extremely successful language for talking about a wide variety of phenomena. Without being dogmatic, one may therefore set aside giving up any of these as implausi-ble, as some of the remaining principles concerning properties are clearly less well entrenched in ordinary reasoning, even when focusing just on reasoning about properties.

This leaves four principles specifically concerned with properties: EXTRACTION, TRANSPARENCY, GENERALITY, and SELF-APPLICATION. As seen above, the first three of these principles can be motivated by ele-mentary reasoning about properties, on subject matters of ordinary and scientific concern such as the moons of Jupiter. In contrast, the need for SELF-APPLICATION does not seem to arise naturally in any such context. To motivate it, the highly abstract and distinctly logical or philosophical case of applying self-identity to itself was invoked.

SELF-APPLICATION therefore stands out as the desideratum whose omission will be least disruptive to naive reasoning about properties. Are the remaining desiderata consistent? Can a language be introduced which satisfies all of them? And will such a language improve on talk of properties in natural language? In the remainder, a case will be made for a positive answer to these questions.

Before moving on to the positive proposal, it is worth acknowledg-ing two limitations to the argument against SELF-APPLICATION. First, one might note that the case of self-identity being applicable to itself does not require the full strength of SELF-APPLICATION: it may be that only some property terms, like $[x.x = x]$ can be applied to themselves, while others, like $[\chi.\neg\chi@\chi]$ in the argument just given cannot. One might therefore prefer to weaken SELF-APPLICATION rather than reject it outright. Second, a more general version of SELF-APPLICATION can be motivated using examples which are less theoretical than the case of self-application. The more general principle says that for any property terms $\rho_1, \ldots, \rho_n$ and n-ary property term π, $\pi@\rho_1 \ldots \rho_n$ is a formula. To motivate it, one might appeal to cases like Socrates loving love, which one might regiment using a formula of the form $l@sl$. Although this case is still a more unusual case of talking about properties than the very mundane cases which motivated the other desiderata, it shows that the cases motivating SELF-APPLICATION need not all be of such a highly theoretical character as self-identity applying to itself.

These two observations bring out that the route which we will follow here, of setting SELF-APPLICATION aside and seeing where the remaining desiderata lead us, is not the only possible route one might want to explore. But there is also no need—and maybe there was never hope— to argue for such a strong conclusion. The aim here has been merely to motivate *a* natural way of constructing a language which improves upon natural language property talk. We have seen no reason to suppose that there is only one useful way of doing so. We return to this point at the end of the book, in section 10.2.

1.4 Metasemantics

The next section introduces a higher-order language $\mathcal{L}^*$ which satisfies the desiderata identified above. The aim will not merely be to describe the syntax of the language. Rather, we will aim to introduce the language in such a way that its expressions are *meaningful*, relative to an interpretation of any non-logical constants and free variables. By specifying an intended interpretation of the language, we ensure that its sentences can be used in assertions, and in general to replace English in our philosophical theorizing.

It will be helpful to consider how the intended interpretation of a familiar logical language may be fixed, before employing any such metasemantic devices in endowing a novel formal language with an intended interpretation. We therefore consider first the case of first-order predicate logic. So, let $\mathcal{L}_1$ be a standard first-order language, built up from individual variables and individual constants, n-ary predicate constants and the identity predicate $=$ using the Boolean connectives $\neg$, $\wedge$, $\vee$, $\rightarrow$, and $\leftrightarrow$, and the quantifiers $\forall$ and $\exists$ binding individual variables. We assume that $\mathcal{L}_1$ has an intended interpretation, which renders every one of its formulas meaningful, relative to an interpretation of its non-logical constants and free variables.

How is the intended interpretation of $\mathcal{L}_1$ given? Plausibly, translations into natural language play a part, but do not constitute the interpretation. Consider the conjunctive connective $\wedge$: the meaning of $\wedge$ is clearly constrained by being intended to correspond to the English word "and." Similarly, the semantic function of an atomic sentence of the form Fa is clearly constrained by being intended to be able to correspond to the English sentence "Socrates sits" on a suitable interpretation of F and a.

However, it would be a mistake to think that the interpretation of $\mathcal{L}_1$ is constrained purely by such correlations to natural language sentences. Another important factor is provided by the calculus of classical first-order logic. The principles of classical first-order logic also serve to constrain the interpretation of $\mathcal{L}_1$, in ways which outstrip those provided by correlations with natural languages. For example, the fact that the sentence $\exists x(x = a)$ is derivable in classical first-order logic serves to constrain the admissible interpretations of non-logical constants, as it rules out certain interpretations of individual constants. In particular, since "There is no such thing as Pegasus" is true, a cannot be interpreted in the same way in which "Pegasus" is interpreted.

A third constraint on the intended interpretation of $\mathcal{L}_1$ is arguably imposed by the standard set-theoretic model theory. It is important to be clear that although such a model theory is often called a "(formal) semantics," it does not on its own make a language meaningful: When defining a model theory, we are merely doing set theory. For example, when introducing the notion of "truth in a model," we are merely introducing a symbol such as $\vDash$ to abbreviate a complex set-theoretic condition. Introducing such a symbol does not make it the case that sentences of $\mathcal{L}_1$ are meaningful. (We return to set-theoretic model theory in section 1.5; see also Burgess (2008) on the distinction between model theory and semantics.) However, models can be *used* to constrain the interpretation of a language like $\mathcal{L}_1$. Roughly, the relation $\vDash$ between models and formulas of the language can be said to *model* the relationship of truth on the intended interpretation between reality and the formal language $\mathcal{L}_1$. As in any case of modeling, some features of the model will correspond to features of what is modeled, while others are mere artifacts. For model theory to serve as a metasemantic constraint, some demarcation will have to be made which distinguishes representational from artifactual features.

Do the three constraints mentioned so far determine an intended interpretation of any sentence of $\mathcal{L}_1$ relative to an interpretation of its non-logical constants and free variables? Some might worry that there is one aspect which has not been sufficiently constrained, which is the range of quantification. They might say that for an intended interpretation to be determined, a domain of quantification needs to be specified as well. In the following, we will be looking for an interpretation of quantified languages which does not require such further specification of domains. However, there is a fourth constraint which we will assume suffices to complete the determination of the intended interpretation of quantifiers, which is the world itself. By not constraining the interpretation of quantifiers, they will be understood as unrestricted, so that the limits of what they quantify over is provided by the world, i.e., by what reality provides to be quantified over.

In the following, we will mainly focus on the first two kinds of metasemantic constraints, and set aside discussion of the latter two constraints. The reason for setting aside the last constraint is simple: the constraints on interpretations set by reality are not within our control, and do not require any stipulations to be in place. The reason for setting aside model theory is more particular to the aims of this book. Recall that a central concern in the remainder of this book will be to determine how finely propositions are individuated. Model theories which go beyond the constraints imposed by axiomatic systems generally make non-trivial assumptions about the individuation of propositions. We will see one example of this in section 2.4, which defines a set-theoretic model theory along the lines of well-known possible worlds models for modal logic. Such a model theory will validate a number of controversial claims about the individuation of propositions, and so cannot be used in the metasemantics of our language without presupposing these claims.

1.5 Going Higher Order

This section introduces a higher-order language $\mathcal{L}^*$ which satisfies the desiderata identified above. Given the first four desiderata, the starting point of the introduction of the higher-order language is the language of first-order logic $\mathcal{L}_1$. As discussed in the previous section, we assume that this language has an intended interpretation, on which every formula is (uniquely) meaningful relative to an interpretation of its non-logical constants and variables. More specifically, we assume that on this interpretation, $\mathcal{L}_1$ satisfies the first four constraints, of BASIC PREDICATIONS, TRUTH-FUNCTIONS, QUANTIFIERS, and NON-CONTRADICTION. The task is now to extend the language so as to satisfy also the remaining desiderata.

Consider the second batch of desiderata: EXTRACTION, ATTRIBUTION, and TRANSPARENCY. In order to satisfy them, introduce, for every formula φ of $\mathcal{L}_1$ and pairwise distinct first-order variables $x_1, \ldots, x_n$, where $n > 0$, a complex n-ary *predicate* $\lambda x_1 \ldots x_n.\varphi$. As in the case of the connectives of $\mathcal{L}_1$, the interpretation of this new construction is constrained both by certain correspondences with natural language expressions and some stipulative logical principles. For the first, consider an interpretation on which B expresses being blue and S expresses being square; then $\lambda x.Bx \wedge Sx$ is stipulated to express being blue and square. (Alternatively, and maybe more carefully, this λ-term may be stipulated to be synonymous with "blue and square" or "is blue and square"; more on this in a moment.) Many similar paradigmatic examples can easily be provided along the same lines. The central stipulative logical principle is the following conversion schema:

$$(\lambda x_1 \ldots x_n.\varphi)t_1 \ldots t_n \leftrightarrow \varphi[t_1, \ldots, t_n/x_1, \ldots, x_n]$$

where $t_1, \ldots, t_n$ are individual terms free for $x_1, \ldots, x_n$ in φ. Call the resulting language $\mathcal{L}_1^\lambda$.

Assuming this expansion of $\mathcal{L}_1$ is successful, $\mathcal{L}_1^\lambda$ satisfies EXTRACTION, ATTRIBUTION, and TRANSPARENCY. For $n > 0$, $[x_1 \ldots x_n.\varphi]$ may be taken to be $\lambda x_1 \ldots x_n.\varphi$, and $[x_1 \ldots x_n.\varphi]@t_1 \ldots t_n$ to be $(\lambda x_1 \ldots x_n.\varphi)t_1 \ldots t_n$. For the degenerate case of $n = 0$, $[.\varphi]$ may simply taken to be φ, and $[.\varphi]@$ to be φ again. Then, TRANSPARENCY is true by stipulation. The only remaining desideratum is GENERALITY. This is not satisfied by $\mathcal{L}_1^\lambda$, as the only kind of variables included in $\mathcal{L}_1^\lambda$ are individual variables, which cannot take the position of predicates in $\mathcal{L}_1^\lambda$.

Thus, let $\mathcal{L}_2$ be the extension of $\mathcal{L}_1^\lambda$ by a new kind of variables which take the position of n-ary predicates. Like predicate constants, these are given a fixed arity: for every natural number n, there are variables which take the position of n-ary predicate constants. Call these (n-ary) *second-order variables*. In order to keep QUANTIFIERS satisfied, let $\mathcal{L}_2$ allow these new variables to be bound by universal and existential quantifiers.

The intended interpretation of these new variables and quantifiers of $\mathcal{L}_2$ is determined by three explicit stipulations. First, being blue and square, and similar interpretations of predicates, are naturally called properties. Thus, the quantificational prefix $\forall X$, where X is an n-ary second-order variable can be said to correspond to the English "every n-ary property." The second component to this determination is again provided by stipulative principles of a logical character. As in the case of first-order quantifiers, the relevant principles are the axiomatic principles of elementary quantification theory. In the case of $\forall X$, a third determinant can be provided. The interpretation of $\forall X$ may be constrained by stipulating that it stands to n-ary predicate constants as individual quantification stands to individual constants.

In evaluating these stipulations, it is important to recall that—as in the case of logical constants of first-order logic—the correlation of formal expressions with natural language phrases is not meant to provide a translation manual, nor to ensure an exact fit in interpretation. This is crucial in the case of quantifiers binding second-order variables, for two reasons.

First, as noted at the outset, naive reasoning about properties in English is liable to lead to inconsistency. This may be due to mistaken beliefs about properties, but it may also be due to semantic or metasemantic defects in English property talk, such as indeterminacy or vagueness. The intended gap between the intended interpretation of $\mathcal{L}_2$ and English property talk allows for the hope that any semantic defects of English need not be passed on to $\mathcal{L}_2$. The second and third component of the stipulative introduction of the new quantifiers therefore play an equally important role. The consistency of the stipulated logical principles can be demonstrated using standard set-theoretic methods (see section 2.5), but it is still a substantial claim that the metasemantic constraints imposed here serve to single out a unique intended interpretation of $\mathcal{L}_2$ (as always relative to an interpretation of non-logical constants and free variables).

Second, English quantificational phrases like "every n-ary property" constitute nominal quantification, and so on their own are most naturally regimented using first-order quantifiers. If such English correlates were the only guides to the intended interpretation of second-order quantifiers, they should arguably be understood as restricted first-order quantifiers, which is not intended. The same point applies to the interpretation of λ-terms. If the only guide to the intended interpretation of $\lambda x.Bx \wedge Sx$ was the stipulation to correspond to the English phrase "being blue and square," it would be natural to consider this λ-term to function semantically much more like an individual term rather than a predicate, which is not intended. As a consequence, the intended discrepancy between the various formal constructions and their English correlates may go substantially beyond merely, e.g., resolving ambiguities. The hope is that this does not trivialize their role as metasemantic guides to the interpretation of the introduced formal constructions.

The addition of second-order variables and quantifiers binding them ensures that $\mathcal{L}_2$ satisfies GENERALITY, but it also leads to failures of EXTRACTION: if X is a second-order variable, there is no property term of the form $[X.\varphi]$. Given the above introduction of λ-terms, the solution is obvious: let X be bindable by λ. In general, for any pairwise distinct variables $\xi_1, \ldots, \xi_n$ of any syntactic categories and formula φ of $\mathcal{L}_2$, let there be a term $\lambda\xi_1 \ldots \xi_n.\varphi$. In order to satisfy ATTRIBUTION, allow this to be predicated of arguments $\tau_1, \ldots, \tau_n$ of corresponding syntactic categories. That is, if ξ_i is an individual variable, τ_i may be any individual term; if ξ_i is an n-ary second-order variable, τ_i may be any n-ary predicate. The interpretation of the new terms is determined as before, by correspondence to English phrases and stipulative logical principles. Let $\mathcal{L}_2^\lambda$ be the resulting expansion of $\mathcal{L}_2$.

$\mathcal{L}_2^\lambda$ satisfies EXTRACTION, ATTRIBUTION, and TRANSPARENCY. But it does not satisfy GENERALITY. For example, if X is a second-order variable and φ is a formula, then $\lambda X.\varphi$ is a property term of $\mathcal{L}_2^\lambda$, but it cannot be replaced by any variable. Thus, $\mathcal{L}_2^\lambda$ suffers from the same kind of limitation from which $\mathcal{L}_1^\lambda$ suffers, just as $\mathcal{L}_2$ suffers from the same kind of limitation from which $\mathcal{L}_1$ suffers. The iterative expansions of languages discussed here can clearly be carried further, leading to an infinite series of pairs of languages

$$\mathcal{L}_1, \mathcal{L}_1^\lambda; \mathcal{L}_2, \mathcal{L}_2^\lambda; \mathcal{L}_3, \mathcal{L}_3^\lambda; \ldots$$

It is not hard to see that no language in this sequence satisfies all desiderata: for any n, $\mathcal{L}_n$ will fail EXTRACTION (and so also ATTRIBUTION), while $\mathcal{L}_n^\lambda$ will fail GENERALITY.

It may seem that the approach chosen here goes round in a circle. But every step in the sequence of languages described here corresponds to a genuine expansion. It is therefore not so much that the process goes round in a circle, but that it goes up a helix. At no point on the helix are all desiderata satisfied. However, let $\mathcal{L}^*$ be the language whose sentences are the sentences of all the languages in this sequence. This language plausibly satisfies all eight desiderata. The crucial cases to consider are EXTRACTION and GENERALITY. Take GENERALITY: any property term must be a term of $\mathcal{L}_i^\lambda$, for some i, whence $\mathcal{L}_{i+1}$ provides a variable which replaces it. The case of EXTRACTION is not much more complicated: any variables $\xi_1, \ldots, \xi_n$ must be variables of some languages $\mathcal{L}_{i_1}, \ldots, \mathcal{L}_{i_n}$. The relevant property term is therefore provided by $\mathcal{L}^\lambda_{\max\{i_j : j \leq n\}}$.

The remaining desiderata are also plausibly satisfied. The cases of ATTRIBUTION, BASIC PREDICATIONS, TRUTH-FUNCTIONS, and QUANTIFIERS are immediate. TRANSPARENCY is plausibly ensured by the introduction of λ-terms. This leaves NON-CONTRADICTION. Unless the stipulative logical principles used in the various extensions are jointly inconsistent, there is no reason to think that NON-CONTRADICTION fails. And the consistency of these principles is easily demonstrated: $\mathcal{L}^*$ is a language of higher-order logic, or relational type theory, which can be given a standard set-theoretic model theory, which provides a relative consistency proof.

$\mathcal{L}^*$ is a version of what is often called *relational* type theory, as opposed to the *functional* type theory of Church (1940). Relational type theories can be seen as simplifications of the *ramified* type theory of Whitehead and Russell (1910–1913); early formulations were given by Carnap (1929), Hilbert and Ackermann (1938), Myhill (1958), and Orey (1959); see also Gallin (1975) for a modal extension.

Higher-order languages face certain important objections, which it will be important to address. However, this is best done once a substantial theory of propositional granularity is argued for. One might also worry that the metasemantic stipulations made here allow for more than one interpretation, and so that further stipulations are required to arrive at a unique intended interpretation. We will in fact encounter one example of such a possible further metasemantic stipulation in section 4.6. For most of this book, it will be assumed that a unique intended interpretation is isolated, whether the stipulations identified in this chapter suffice for this or not. We will return to these methodological issues in the two chapters of the concluding Part V.

2

A Formal Language

The previous chapter motivated adopting a formal language to improve talk of entities like propositions, properties, relations, and modalities, and argued for the language $\mathcal{L}^*$ of relational type theory. The remainder of this book will be concerned in particular with propositions and modalities. For much of the following, a fragment of $\mathcal{L}^*$ will therefore suffice, namely the one which includes only sentential expressions (which express propositions) and sentential operators (which express modalities). This chapter introduces this language more rigorously.

Working with such a restricted language is not essential; all of the following could be carried out in a richer language such as $\mathcal{L}^*$. But the restricted language will simplify a number of matters. Further, it will turn out that certain additional resources need to be appealed to, which are not even included in $\mathcal{L}^*$. It is therefore inadvisable to attempt to find a perfect language in which all theorizing can be regimented; even if such a language exists, it would likely be impractically complex. Instead, it is best to adopt a flexible approach, and work with different languages depending on the purpose at hand. For much of the following, it is best to work with the relatively limited fragment of $\mathcal{L}^*$ which will be introduced in this chapter.

Before proceeding to a rigorous statement of the syntax of this fragment, it is worth being clear about how the words "proposition" and "modality" will be used in the following. Since $\mathcal{L}^*$ is meant to improve on ordinary uses of these words, they are meant to be used in a special technical sense here. This technical sense is provided by $\mathcal{L}^*$, in the sense that the interpretation of discussions involving "proposition" and "modality" in the following is meant to be tied to the intended interpretation of $\mathcal{L}^*$, as introduced above. Ordinary uses of English words like "proposition" and "modality" played a role in determining this intended interpretation,

The Foundations of Modality: From Propositions to Possible Worlds. Peter Fritz, Oxford University Press.
© Peter Fritz 2023. DOI: 10.1093/oso/9780192870025.003.0003

but so did other constraints; the present use of these words may therefore likely come apart from their ordinary use.

One example of this deviation is the present use of "modality": I will use this term for the kind of entities which are expressed by sentential operators, and so what quantifiers binding variables in the position of sentential operators range over. Since truth-functional connectives are sentential operators, this means that I count, e.g., negation as a (truth-functional) modality. This wide usage of "modality" is not meant to indicate a philosophically substantial thesis; in the following, a term like "property of propositions" could as well be used instead.

2.1 Syntax

The syntactic categories, called *types*, of the language $\mathcal{L}$ to be introduced consist of the type of formulas (sentential terms), and the type of operators taking n formulas as arguments. An n-ary operator takes n formulas as arguments to produce a formula. A formula can therefore be considered to be the limiting case of an operator, taking no formulas as arguments to produce a formula; thus, formulas can be considered as nullary operators. Consequently, the types of $\mathcal{L}$ can be taken to be the natural numbers, where any $n \in \mathbb{N}$ is the type of n-ary operators. Formulas are then simply expressions of type 0. (Since the type of an expression can be thought of as its arity, this number is therefore entirely unrelated to the height of the different levels in the hierarchy of expressions of the more general higher-order language $\mathcal{L}^*$ discussed in the previous chapter.)

The syntax of $\mathcal{L}$ is defined in the usual recursive way, specifying for each type n the set of expressions of type n. The language is based on a set of variables and constants for each type, which are assumed to be pairwise disjoint. The number of variables of each type is countably infinite. The choice of constants is left somewhat open at this point. This might be considered a parameter of the construction, but one that will be left implicit for simplicity. Naturally, variables and constants of some type are expressions of that type. Formulas are closed under the usual Boolean connectives and universal and existential quantifiers binding any variables. Formulas can also be constructed by applying an

n-ary operator to n formulas. Finally, a complex n-ary operator can be constructed from any formula using λ-abstraction, employing any non-empty sequence of n pairwise distinct sentential variables. This leads to the following definition:

Definition 2.1. *Let $\mathcal{L}$ be the smallest language which provides, for every type $t \in \mathbb{N}$, a set of expressions such that:*

(1) *Every variable and constant of type t is an expression of type t.*

(2) *If φ and ψ are expressions of type 0, then $(\neg\varphi)$, $(\varphi \wedge \psi)$, $(\varphi \vee \psi)$, $(\varphi \rightarrow \psi)$, and $(\varphi \leftrightarrow \psi)$ are expressions of type 0.*

(3) *If φ is an expression of type 0 and x is a variable, then $(\forall x\varphi)$ and $(\exists x\varphi)$ are expressions of type 0.*

(4) *If μ is an expression of type $n > 0$ and $\varphi_1, \ldots, \varphi_n$ are expressions of type 0, then $(\mu\varphi_1 \ldots \varphi_n)$ is an expression of type 0.*

(5) *If φ is an expression of type 0, $n > 0$, and $p_1, \ldots, p_n$ are pairwise distinct variables of type 0, then $(\lambda p_1 \ldots p_n.\varphi)$ is an expression of type n.*

A formula is an expression of type 0. A sentence is a formula without any free variables. For $n \in \mathbb{N}$, an n-ary operator is an expression of type n. A propositional variable is a variable of type 0.

The definition of *free* and *bound* (occurrences of) variables is standard, with $\forall$, $\exists$, and λ binding variables. As usual, a variable x occurring in an expression ε is *free for* an expression η of the same type if in the result of replacing every free occurrence of x in ε by η, no occurrence of a variable free in η becomes bound. In this case, this resulting expression is notated $\varepsilon[\eta/x]$. This is also written $\varepsilon(\eta)$ when x is contextually salient or unspecified. We typically use $p, q, r, \ldots$ for propositional variables, and $m, n, o, \ldots$ for operator variables of positive arity. We typically use $\varphi, \psi, \ldots$ for arbitrary formulas, and $\mu, \nu, \ldots$ for arbitrary operators.

Brackets will be omitted when this does not introduce ambiguities. In order to facilitate this, the usual conventions will be assumed, on which unary operators bind more strongly than binary operators, and $\wedge$ and $\vee$ bind more strongly than $\rightarrow$ and $\leftrightarrow$. For example, using $\Box$ as a constant of type (arity) 1, $((\Box(p \wedge q)) \rightarrow ((\Box p) \wedge (\Box q)))$ can be written as $\Box(p \wedge q) \rightarrow \Box p \wedge \Box q$. Further, a finite sequence will be indicated using a bar over the relevant expression. For example, $\lambda p_1 \ldots p_n.\varphi$ will be abbreviated to $\lambda \bar{p}.\varphi$. Similarly, a sequence of quantifiers $\forall x_1 \ldots \forall x_n$ is indicated using $\forall \bar{x}$.

To see how $\mathcal{L}$ can be considered to be a fragment of $\mathcal{L}^*$, note that propositional variables (i.e., variables of type 0) can be understood as nullary second-order variables. Variables of type $n \geq 0$ can be understood as n-ary third-order variables, taking n formulas as arguments. Informally, they can be said to stand for n-ary properties of propositions (i.e., nullary properties), or in other words, n-ary modalities. $\mathcal{L}$ can therefore be seen as a particular fragment of the language of third-order logic (itself a fragment of the more general higher-order language $\mathcal{L}^*$), which omits all expressions involving individual terms in any way.

Two aspects of the treatment of logical connectives are worth discussing. First, all of the usual Boolean connectives and quantifiers are introduced as primitive syntactic constructions. In many logical systems, a proper subset is taken as primitive, and the remaining ones are introduced as abbreviations. This is often unproblematic, as the relevant systems allow for truth-conditionally equivalent formulas to be replaced in any context. In the present context, it is important not to assume this, as doing so would prejudge a number of questions about the individuation of propositions which will be central in the following (see especially the next two chapters). Second, Boolean connectives could be taken as unary and binary operator constants, rather than treated in the present syncategorematic way. The present treatment is not due to any interest in a distinction between logical and non-logical constants, but because it will later be useful to consider fragments of $\mathcal{L}$ which omit formulas constructed by clause (4), while still being closed under Boolean connectives.

In this section, the expressions of $\mathcal{L}$ are *mentioned*. As usual, no quotation marks are used to indicate this, since in the context of an English sentence, any occurrence of an expression of $\mathcal{L}$ is most naturally interpreted as a singular term referring to the relevant expression. However, since $\mathcal{L}$ has been endowed with an intended interpretation (relative to an interpretation of the non-logical constants), a sentence of $\mathcal{L}$ can also be *used* to make an assertion. Context will disambiguate between use and mention. As usual, no distinction will be drawn between asserting a claim, and asserting the truth of the claim. The view which we will ultimately end up with here vindicates this practice.

2.2 A Proof System

In the previous chapter, $\mathcal{L}^*$ was introduced using certain principles of a logical character to constrain the meanings of its expressions. These included principles capturing the classical truth-functional behavior of Boolean connectives, classical principles of elementary quantification theory, and a standard principle of (material) λ-conversion (as formulated in the desideratum TRANSPARENCY). These principles may therefore be considered to be true by virtue of the way in which $\mathcal{L}^*$ was introduced. The corresponding claim thus also holds for its fragment $\mathcal{L}$. It will be useful to have a rigorous way of stating this assumption. This most naturally takes the form of a proof system.

The proof system consists of a number of axioms and rules, corresponding to the logical principles just described. There are a number of ways of constructing such a proof system, corresponding to different ways of encoding the classical principles of Boolean connectives and quantifiers. In the following, it will be useful to employ a Hilbert calculus. A Hilbert calculus is a proof system based on a choice of axioms and rules, with a formula φ being derivable just in case there is a proof of φ from the axioms using the rules. Here, a proof is a finite sequence of formulas each of which is either an axiom or a formula which can be derived from some of the foregoing formulas in the sequence using one of the rules; a proof of φ is a proof which contains φ. To define our proof system, let a *tautology* be a formula consisting only of propositional variables and Boolean connectives which is a truth-functional tautology in the usual sense. This can be defined more rigorously using truth-tables, which provides a simple decision procedure. The set of tautologies can therefore simply be taken as an axiom schema. With this, the proof system is defined as follows:

Definition 2.2. $\vdash$ *is the proof system consisting of the following axiom schemas and rules in $\mathcal{L}$:*

(TAUT)	*tautologies*	(MP)	$\varphi, \varphi \to \psi / \psi$
(UI)	$\forall x\varphi \to \varphi[\varepsilon/x]$	(UG)	$\varphi / \forall x\varphi$
(UE)	$\exists x\varphi \leftrightarrow \neg\forall x\neg\varphi$	(λC)	$(\lambda\bar{p}.\psi)\bar{\varphi} \leftrightarrow \psi[\bar{\varphi}/\bar{p}]$
(UD)	$\forall x(\varphi \to \psi) \to (\varphi \to \forall x\psi)$		
	(x not free in φ)		

φ *being derivable in $\vdash$ is written as $\vdash \varphi$.*

The axioms of this system are either tautologies, principles of elementary quantification theory (UI, UD, and UE), or λ-conversion. On any interpretation of the variables and constants, these may therefore be assumed to be true. Similarly, by the truth-functional behavior of $\rightarrow$, MP is guaranteed to preserve truth under any interpretation of the variables and constants. Finally, by the intended interpretation of the universal quantifier, if φ is true under every interpretation of the variables and constants, then so is $\forall x\varphi$. By induction on the length of proofs, it follows that if $\vdash \varphi$, then φ is true under every interpretation of the variables and constants.

Moreover, the axioms and rules of $\vdash$ correspond to standard Hilbert axiom systems of classical first-order logic. For every derivation in such a proof system, there is a corresponding derivation in $\vdash$. In this sense, $\vdash$ captures all of elementary quantification theory. Elementary quantification theory does not contain λ, and the interaction between the quantifier principles and λC leads to some important consequences. Central among them is a comprehension principle, according to which there is a modality corresponding to any condition. This is the following schematic principle, which can be instantiated using any operator variable o and formula φ which does not contain any free occurrence of o:

(COMP) $\exists o \forall \bar{p}(o\bar{p} \leftrightarrow \varphi)$

Lemma 2.3. $\vdash$ COMP.

Proof. Contraposing an instance of UI, the following is derivable:

$$\forall \bar{p}((\lambda \bar{p}.\varphi)\bar{p} \leftrightarrow \varphi) \rightarrow \exists o \forall \bar{p}(o\bar{p} \leftrightarrow \varphi)$$

The antecedent can be derived by applying UG to an instance of λC.   □

Given the introduction of $\mathcal{L}^*$, the proof system $\vdash$ is a very natural starting point in the restricted language $\mathcal{L}$: it allows the derivation of just those formulas whose truth (under any interpretation of the variables and constants) follows directly from the stipulations made in the introduction of these formal languages. As will shortly be seen, there are formulas which are true under any interpretation of the variables and constants but which are not derivable in $\vdash$. In this sense, $\vdash$ is incomplete with respect to the intended interpretation of $\mathcal{L}$. But such completeness was not the aim of the introduction of ⊢; indeed, we will see later that such completeness may well be unattainable. Rather, ⊢ is meant to make a start at codifying a useful body of truths on which the following will build. It may be worth stressing that there is no assumption that the theorems of ⊢ are *logical* truths, in any substantial sense of the word; such a notion of logicality will be neither needed nor employed in the following.

In exploring various views and their consequences, it will be useful to be able to notate concisely when one claim follows from another, given the commitments codified in φ. In the case of sentences, φ following from ψ, given ⊢, is naturally understood as the conditional claim $\psi \to \varphi$ being derivable in ⊢; i.e., ⊢ $\psi \to \varphi$. Two generalizations of this notion will be useful as well. The first allows for multiple premises. This can be accommodated by letting φ follow from a set Γ if ⊢ $\gamma_1 \wedge \ldots \wedge \gamma_n \to \varphi$, for some $\gamma_1, \ldots, \gamma_n \in \Gamma$. The second allows for the relevant formulas to contain free variables. In this case, it will be useful to take free variables to indicate implicit universal quantifiers. This simplifies the statement of various principles formulated in $\mathcal{L}$, since any initial universal quantifiers may simply be dropped. A set of formulas Γ being inconsistent can be defined as Γ allowing the derivation of some formula whose negation is derivable in ⊢; here, the formula $\exists p(p \wedge \neg p)$ is chosen by convention.

Definition 2.4. *For any formula ψ, let $\bar{\forall}\psi$ be the sentence $\forall x_1 \ldots \forall x_n\psi$, where $x_1, \ldots, x_n$ are the variables free in ψ (according to some fixed ordering).*

For any formula φ and set of formulas Γ, let φ follow from Γ in ⊢, written $\Gamma \vdash \varphi$, if ⊢ $\bar{\forall}\gamma_1 \wedge \ldots \wedge \bar{\forall}\gamma_n \to \bar{\forall}\varphi$ for some $\gamma_1, \ldots, \gamma_n \in \Gamma$.

Define:

$$\bot := \exists p(p \wedge \neg p)$$
$$\top := \neg\bot$$

A set of formulas Γ is inconsistent *in* ⊢ *if $\Gamma \vdash \bot$, and* consistent *in* ⊢ *otherwise.*

When making use of these notions, the set Γ will often be indicated simply by listing its elements. In particular when Γ has one member γ, this formula will be used to stand for its singleton $\{\gamma\}$. We also specify the extension of a set Γ by a formula φ using +, writing $\Gamma + \varphi$ for $\Gamma \cup \{\varphi\}$.

The qualifier "in ⊢" will sometimes be dropped when this is clear from context.

Languages like $\mathcal{L}$ are uncommon in contemporary philosophical logic, but they have a significant history. An important example of a system in such a language is the *protothetic* of Leśniewski (1929), which—following results in the doctoral thesis of Tajtelbaum-Tarski (1923)—incorporates quantifiers binding variables in the position of unary sentential operators. For introductions to this system, see Słupecki (1953), Grzegorczyk (1955), and Simons (2020); see also Henkin (1963) and Andrews (1963) for higher-order extensions of it. All of these deductive systems assume principles of extensionality, according to which p and q are interchangeable when materially equivalent. As Church (1956, §28) notes, this renders all of the quantifiers eliminable along the lines discussed in Chapter 4. However, corresponding systems without such an extensionality principle were investigated as well, e.g., by Prior (1955, 1957) and Cresswell (1966, 1967, 1972).

2.3 Identity

Identity among propositions will be a crucial topic in the discussion below. To regiment this, it will be useful to assume that the language under consideration includes a binary constant $=$ whose intended interpretation is propositional identity. $\mathcal{L}^=$ will be used to stand for the language $\mathcal{L}$ under this assumption. More carefully, the intended interpretation of $=$ can be determined in ways which are analogous to the determination of the interpretation of Boolean connectives and quantifiers above. That is, the interpretation of $=$ is narrowed down, first, by the standard axiomatic principles of identity, second, by stipulating that it corresponds (but most likely is not identical) to what is expressed by the English phrase "is a proposition identical to," and third, by stipulating that it stands to $=$ as used in first-order logic as propositional variables stand to first-order variables. As with the axioms of ⊢, the axiomatic principles of identity may therefore be taken to be true by virtue of the introduction of $=$. Along the lines of standard axiomatic systems including identity, these principles may be taken to consist of an axiom stating the reflexivity of identity, and an axiom stating Leibniz's Law in the form that identical propositions share the same properties (modalities). For familiarity, $=$—along with other binary operators—will be notated infix; that is, a formula of the form $=\varphi\psi$ will also be written $\varphi = \psi$. Likewise, the negation $\neg(\varphi = \psi)$ of such a formula will be abbreviated as $\varphi \neq \psi$.

Definition 2.5. *Let* ID *be* {RI, LL}, *where:*

(RI) $p = p$
(LL) $p = q \rightarrow (op \leftrightarrow oq)$

$\vdash^= \varphi$ *will be written for* ID $\vdash \varphi$; *further notions such as* following from *are derived as in the case of* $\vdash$.

Two variants of Leibniz's Law can be derived from these principles along standard lines. One of them states that propositions have the same properties if and only if they are identical. The other states, schematically, that $\varphi(p)$ is $\varphi(q)$ whenever $p = q$:

Lemma 2.6. *The following are derivable in* $\vdash^=$:

$$p = q \leftrightarrow \forall o(op \leftrightarrow oq)$$
$$p = q \rightarrow (\varphi(p) = \varphi(q))$$

Proof. Routine.   □

No connective for identity among modalities will be included in the following, mainly for simplicity. However, using the resources of $\mathcal{L}$, one can define a connective $\equiv$ which expresses that modalities are functionally equivalent, in the following sense: μ and ν are functionally equivalent if for every proposition p, the proposition μp is the proposition νp. For uniformity, it will be useful to extend this notion of functional equivalence to propositions, in which case functional equivalence may be identified with identity. That is, let $\equiv$ serve to indicate formulas as follows:

$$\mu \equiv \nu := \forall \bar{p}(\mu\bar{p} = \nu\bar{p})$$
$$\varphi \equiv \psi := \varphi = \psi$$

The axiomatic principles of $\vdash^=$ say little about what it takes for propositions to be identical. It is thus not to be expected that it can be proven, e.g., that materially equivalent propositions are identical. And indeed, this principle cannot be derived. To show such facts about what cannot be derived, it will be useful to introduce models. This will be done in the final, supplementary, two sections of this chapter.

2.4 Supplement: Possible Worlds Models

For many purposes in the following, a familiar kind of model will suffice. These models are based on the idea of possible worlds. Each model specifies a set of elements which are informally considered as possible worlds. Possible worlds determine the truth or falsity of each formula of $\mathcal{L}$, relative to a valuation function used to interpret constants, and an assignment function used to interpret variables.

In these possible worlds models, a proposition is modeled as the set of worlds in which it is true. This imposes some strong constraints on the individuation of propositions, which have not been argued for. Although this limits the applicability of the models, they will turn out to be flexible enough for many purposes. Similar to propositions, modalities are individuated modally in these models: an n-ary modality will be modeled as a function from n-tuples of propositions to propositions. For example, a binary modality of necessary equivalence will be modeled as the function which maps any two propositions to the proposition that they are necessarily equivalent.

These basic ideas suggest the following simple construction of domains of quantification (which will be refined in a moment): First, the domain of quantifiers binding variables of type 0 is the set of subsets of the set of worlds W. The set of subsets of a given set X is called the *powerset of* X, and written $\mathcal{P}(X)$. For brevity, let P_0 be $\mathcal{P}(W)$. Second, the domain of quantifiers binding variables of some type $n > 0$ is the set of functions from the set of n-tuples of elements of P_0 to P_0. The set of functions from a given set X to a given set Y will be written as Y^X. Assuming that $n = \{0, \ldots, n-1\}$ for every $n \in \mathbb{N}$, X^n is the set of functions mapping the first n natural numbers to elements of X; these can be identified with the n-tuples on X. P_0^n will thus be considered as the set of n-tuples of propositions, whence the domain of n-ary modalities can be written as $P_0^{P_0^n}$. This will be abbreviated as P_n.

To obtain models which are a bit more flexible, it will be useful to generalize this simple construction of quantificational domains in two ways. First, the domains of quantification will be allowed to be incomplete, in the sense of being merely a subset of P_n, for the relevant type n. Second, the domain of quantification will be allowed to depend on the world of evaluation. So, models will specify a domain function, which assigns to every world w and type n a domain $D_n^w \subseteq P_n$, over which quantifiers binding variables of type n range at world w.

Constants can now be interpreted using a valuation function which maps any constant of type n to an element of P_n, and variables similarly using an assignment function which maps any variable of type n to an element of P_n. A model can then be defined as constituted by a set of worlds W, a domain function D, and a valuation function V. Relative to an assignment function a, every such model determines the interpretation of any expression ε of type n. This is an element of P_n, and will be written as $[\![\varepsilon]\!]_a$. The interpretation of constants and variables are straightforwardly determined by the valuation and assignment functions, respectively. The interpretation of complex expressions is determined recursively. For example, the proposition expressed by a negated formula $\neg\varphi$ is determined by the interpretation of φ; since $\neg\varphi$ is true in a world w just in case φ is not true in w, $[\![\neg\varphi]\!]_a$ is the relative complement of $[\![\varphi]\!]_a$, i.e., the set of worlds not contained in $[\![\varphi]\!]_a$. The interpretation of formulas constructed using other Boolean connectives is determined similarly. Quantified formulas are interpreted as suggested by the construction of D: for example, $[\![\forall x\varphi]\!]_a$ is the set of worlds w in which φ is true under any variant of a which assigns to x any value in D_n^w, where n is the type of x. To specify such variant functions, the following conventions will be adopted. A function f mapping members of X to members of Y will be notated as $f : X \to Y$. For any such function f, $x \in X$ and $y \in Y$, $f[y/x]$ is the function mapping x to y and every other x' to $f(x')$; this is extended to sequences by letting $f[\bar{y}/\bar{x}]$ be $f[y_1/x_1] \ldots [y_n/x_n]$.

All that remains is to specify the interpretation of expressions constructed using the application of an operator or λ-abstraction. The first is straightforward. For a formula of the form $\mu\varphi_1 \ldots \varphi_n$, μ is interpreted as a function from P_0^n to P_0, and $\varphi_1, \ldots, \varphi_n$ are interpreted as elements of

P_0. $\mu\varphi_1 \ldots \varphi_n$ can thus be interpreted as the value obtained by applying this function to the n-tuple of these arguments. Correspondingly, an abstract of the form $\lambda\bar{p}.\varphi$ is of type n (where n is the length of $\bar{p}$), so it must be interpreted as a function from P_0^n to P_0. Given an n-tuple $\bar{d}$, the interpretation of φ when variables $\bar{p}$ are interpreted as propositions $\bar{d}$ can be written as $[\![\varphi]\!]_{a[\bar{d}/\bar{p}]}$. Consequently, $\lambda\bar{p}.\varphi$ will be interpreted as the function which maps any n-tuple $\bar{d} \in P_0^n$ to $[\![\varphi]\!]_{a[\bar{d}/\bar{p}]}$. This function will be written as $\bar{d} \in P_0^n \mapsto [\![\varphi]\!]_{a[\bar{d}/\bar{p}]}$, using the convention to write $x \in X \mapsto \alpha(x)$ for the function mapping every x in a given set X to $\alpha(x)$.

The following definition sums up this model construction more formally, and introduces the symbol $\vDash$ which will be used to state facts about truth and validity in a model. Due to the fact that domains of quantification are variable, it will be helpful to use a slightly non-standard notion of validity, according to which a formula is valid on a model just in case it is true in every world of the model under any assignment function which maps every variable to an element of the relevant domain associated with this world.

Definition 2.7. *A possible worlds model is a structure* $\mathfrak{M} = \langle W, D, V \rangle$, *satisfying the following requirements. First, W is a set. For every $n > 0$, let:*

$$P_0 := \mathcal{P}(W)$$

$$P_n := P_0^{P_0^n}$$

Second, for every type $n \in \mathbb{N}$ and $w \in W$:

$$D_n^w \subseteq P_n$$

Third, V is a function mapping every constant of type n to an element of P_n.

An assignment function is a function mapping every variable of type n to an element of P_n. Define, relative to such an assignment function a, a function $[\![\cdot]\!]_a$ mapping every expression of $\mathcal{L}$ of type n to a member of P_n, as follows:

$$[\![x]\!]_a = a(x)$$

$$[\![c]\!]_a = V(c)$$

$$[\![\neg\varphi]\!]_a = W \backslash [\![\varphi]\!]_a$$

$$[\![\varphi \wedge \psi]\!]_a = [\![\varphi]\!]_a \cap [\![\psi]\!]_a$$

$$\llbracket \varphi \vee \psi \rrbracket_a = \llbracket \varphi \rrbracket_a \cup \llbracket \psi \rrbracket_a$$

$$\llbracket \varphi \rightarrow \psi \rrbracket_a = (W \backslash \llbracket \varphi \rrbracket_a) \cup \llbracket \psi \rrbracket_a$$

$$\llbracket \varphi \leftrightarrow \psi \rrbracket_a = (\llbracket \varphi \rrbracket_a \cap \llbracket \psi \rrbracket_a) \cup ((W \backslash \llbracket \varphi \rrbracket_a) \cap (W \backslash \llbracket \psi \rrbracket_a))$$

$$\llbracket \forall x \varphi \rrbracket_a = \{w \in W : w \in \llbracket \varphi \rrbracket_{a[d/x]} \text{ for all } d \in D_n^w\}$$

$$\llbracket \exists x \varphi \rrbracket_a = \{w \in W : w \in \llbracket \varphi \rrbracket_{a[d/x]} \text{ for some } d \in D_n^w\}$$

$$\llbracket \mu \varphi_1 \ldots \varphi_n \rrbracket_a = \llbracket \mu \rrbracket_a (\langle \llbracket \varphi_1 \rrbracket_a, \ldots, \llbracket \varphi_n \rrbracket_a \rangle)$$

$$\llbracket \lambda \bar{p}.\varphi \rrbracket_a = \bar{d} \in P_0^n \mapsto \llbracket \varphi \rrbracket_{a[\bar{d}/\bar{p}]}$$

A formula φ is true *relative to a possible worlds model $\mathfrak{M}$, world $w \in W$, and assignment function a, written $\mathfrak{M}, w, a \vDash \varphi$, if $w \in \llbracket \varphi \rrbracket_a$.*

For every world $w \in W$, a w-assignment is a function mapping every variable of type n to an element of D_n^w.

φ *is w-valid if $\mathfrak{M}, w, a \vDash \varphi$ for every w-assignment a.*

φ *is* valid on $\mathfrak{M}$ *if φ is w-valid for every $w \in W$.*

The main purpose of models in this book is to establish consistency proofs in a standard way. Roughly, if all formulas of some given set are valid on a model which validates the theorems of a given proof system, then this set is consistent in this proof system. The following lemma states this claim more precisely, using just the notion of w-validity:

Lemma 2.8. *If there is a possible worlds model $\mathfrak{M}$ containing a world w such that:*

(1) *every theorem of a given proof system $\vdash^*$ is w-valid, and*

(2) *every member of a given set of formulas Γ is w-valid,*

then Γ is consistent in $\vdash^*$.

Proof. Assume $\mathfrak{M}$ and w satisfy (1) and (2). For every $\gamma \in \Gamma$, since γ is w-valid, so is $\bar{\forall}\gamma$. Since $\neg\bot$ is w-valid, it follows that there are no $\gamma_1, \ldots, \gamma_n \in \Gamma$ for which $\bar{\forall}\gamma_1 \wedge \ldots \wedge \bar{\forall}\gamma_n \rightarrow \bot$ is w-valid. By (1), there is therefore no conditional of this form which is a theorem of $\vdash^*$, which means that Γ is consistent in $\vdash^*$. □

To establish consistency results in $\vdash$, it would be ideal to be able to prove that every theorem of $\vdash$ is valid on every model. But possible worlds models as defined here are too flexible to allow this: since the domain function is completely unconstrained, UI is not guaranteed to be valid. However, all the other axioms of $\vdash$ are valid on every model, and the rules of $\vdash$ preserve validity. In fact, it will be useful to establish that the rules preserve both w-validity, for a given world w, and validity. To do so, it is helpful to establish first a preliminary lemma on assignment functions and replacements:

Lemma 2.9. *Let ε be an expression of some type $n \in \mathbb{N}$. Then in every possible worlds model:*
  (1) *If a and b are assignment functions such that $a(x) = b(x)$ for every variable x free in ε, then $[\![\varepsilon]\!]_a = [\![\varepsilon]\!]_b$.*
  (2) *If η is free for x in ε, then $[\![\varepsilon]\!]_{a[[\![\eta]\!]_a/x]} = [\![\varepsilon[\eta/x]]\!]_a$.*

Proof. Each claim can be established by a routine induction on the complexity of ε, with the argument for claim (2) relying on claim (1). $\quad\square$

Lemma 2.10. *Let $\mathfrak{M}$ be a possible worlds model.* TAUT, UD, UE, *and* λC *are valid on $\mathfrak{M}$.* MP *and* UG *preserve w-validity for a given world w, and validity on $\mathfrak{M}$.*

Proof. Routine, using Lemma 2.9 (2) for the case of λC. $\quad\square$

Although UI is not valid on all possible worlds models, it is valid on all possible worlds models in which the domain function is unrestricted. Similarly, since the interpretation of $=$ is left unspecified in possible worlds models, RI and LL are not valid on all of them, although these axioms are valid on models which interpret identity such that an identity statement is true in a world just in case the formulas flanking the identity sign are interpreted the same. It will be useful to introduce terminology for these constraints, and to prove these validity claims:

Definition 2.11. *Let $\mathfrak{M} = \langle W, D, V \rangle$ be a possible worlds model.*
  $\mathfrak{M}$ *is* full *if $D_n^w = P_n$ for every type $n \in \mathbb{N}$ and world $w \in W$.*
  $\mathfrak{M}$ *is* simple *if for all $w \in W$ and $d, e \in P_0$: $w \in V(=)(\langle d, e \rangle)$ iff $d = e$.*
  $\mathfrak{M}$ *is* standard *if it is full and simple.*

Lemma 2.12. *Let $\mathfrak{M}$ be a possible worlds model.*
  (1) *If $\mathfrak{M}$ is* full, *then* UI *is valid on $\mathfrak{M}$.*
  (2) *If $\mathfrak{M}$ is* simple, *then* RI *and* LL *are valid on $\mathfrak{M}$.*

Proof. Routine, using Lemma 2.9 (2) for case (1).   □

As a very simple example of a consistency proof, the following result shows that the claim that there are distinct truths is consistent in $\vdash^=$:

Proposition 2.13. $\exists p \exists q(p \wedge q \wedge (p \neq q))$ *is consistent in* $\vdash^=$.

Proof. Let $\mathfrak{M}$ be a standard possible worlds model based on a two-element set $W = \{w, v\}$. By an induction on the length of derivations, it follows from Lemmas 2.10 and 2.12 that every theorem of $\vdash^=$ is valid on $\mathfrak{M}$. Further, the pair of propositions $\{w\}$ and W witnesses the w-validity of $\exists p \exists q(p \wedge q \wedge (p \neq q))$. The claim therefore follows by Lemma 2.8.   □

2.5 Supplement: Relative Consistency

By Lemma 2.8, the consistency of a theory $\Gamma \subseteq \mathcal{L}$ follows from the existence of a suitable model. In the present metaphysical context, it is important to be clear on the status of this claim. Models as defined above are set-theoretic constructs. Someone who prefers higher-order theorizing may be skeptical about a rich ontology of sets, since they may hold that many of the roles played by set theory in contemporary philosophy can be played as well, if not better, by higher-order quantification. They may therefore doubt that there are any sets, and so that there are the required models. Fortunately, this does not mean that model-theoretic consistency proofs are unavailable to them: more carefully, these results can be understood as *relative* consistency proofs, which show that if ZFC set theory is consistent, then Γ is consistent as well. In fact, since ZFC is consistent if ZF is consistent, any such argument shows that if ZF is consistent, then Γ is consistent as well. None of this is specific to the models developed here, but since the ontological status of sets is especially important in the present context, the following will sketch how the relevant relative consistency argument can be carried out.

For purposes of illustration, consider a set of formulas Γ of $\mathcal{L}$. The objective is to show that if ZFC is consistent, then Γ is consistent in $\vdash$. First, the syntax of $\mathcal{L}$ is formulated in ZFC, along with the notion of derivability in $\vdash$. The notions of a model and of a formula being valid on a model are formulated in ZFC. Two facts (formulated in ZFC) are proven in ZFC about these notions. First, that there is a model on which every member of Γ is valid. Second, that every formula derivable in $\vdash$ is valid on every model. Assuming that ZFC proves that models interpret Boolean connectives and quantifiers according to the standard truth-conditions, it follows that ZFC proves the claim (formulated in ZFC) that Γ is consistent in $\vdash$. Next, it is shown by an induction on the length of derivations that if a formula φ is derivable in $\vdash$, then this fact (formalized in ZFC) can be derived in ZFC. Thus, if Γ is in fact inconsistent in $\vdash$, then this fact (formalized in ZFC) can be derived in ZFC. But the negation of this claim was already noted to be derivable in ZFC. So, if Γ is in fact inconsistent, then ZFC is inconsistent. By contraposition, if ZFC is consistent, then Γ is consistent.

The consistency of ZFC cannot be established deductively without relying on the consistency of a theory which is—in a certain sense—stronger than ZFC, due to the incompleteness theorems of Gödel (1931). But there is substantial evidence for the consistency of ZFC. First, there is a vast body of research into the consequences of ZFC, which has not uncovered any inconsistencies. Second, ZFC can be seen as codifying the iterative conception of sets, which spatial imagination shows to be coherent: it has appeared to many set theorists at least to be conceivable that there are objects ordered by a membership relation according to the iterative conception, in a way which validates the axioms of ZFC (and without violating the principles of classical first-order logic). Independently of any questions about whether there is an interpretation of set-theoretic language on which the axioms of ZFC are true, there are therefore strong reasons to think that ZFC is at least consistent.

Assuming then that ZFC is consistent, set-theoretic models can be used to establish consistency facts. This is the use to which models will be put here. As noted above, models *can* play other roles. Here, however, no such further uses of models will be appealed to. In particular, models play no role in determining the intended interpretation of $\mathcal{L}$. In fact, no particular class of models will be singled out here as uniquely important;

instead, various classes of models are constructed *ad hoc* depending on the underivability fact to be established. Finally, no assumption about ZFC will be made except its consistency.

Although in the following, it will simply be assumed that ZFC is consistent, and so that the consistency of various logics and theories can be established using set-theoretic models, many of these consistency results could be established from far weaker assumptions. For example, consider the consistency of the basic axiom system $\vdash$ itself, i.e., the fact that $\nvdash \bot$. As noted above, by adding axioms of extensionality, quantifiers can be eliminated. Along these lines, the consistency of $\vdash$ can be reduced to the consistency of classical propositional logic, which is not in doubt.

PART II
IDENTITY

3
Attitudes and Structured Propositions

Using the formal language introduced in the previous chapter, this chapter begins investigating theories of propositions and modalities. The starting point is identity and the question, how finely propositions are individuated.

3.1 Attitude Ascriptions

We already saw in the Introduction that intuitive judgements about ascriptions of propositional attitudes like belief cannot straightforwardly be used to argue for the distinctness of propositions, since these judgements are not just sensitive to propositions themselves, but also to the way propositions are represented. However, theories of propositional attitude ascriptions often appeal to fine-grained notions of propositions—see, e.g., Salmon (1986) and Soames (1987)—so it is worth considering in more detail how these issues play out in the formal language $\mathcal{L}$ introduced in the previous chapter.

For concreteness, consider the following argument. Let *Peirce's Law* be the following sentence:

(PL) $\forall p \forall q(((p \rightarrow q) \rightarrow p) \rightarrow p)$

Since $((p \rightarrow q) \rightarrow p) \rightarrow p$ is a tautology, PL is derivable in $\vdash$. But that $((p \rightarrow q) \rightarrow p) \rightarrow p$ is a tautology is not immediately obvious, in contrast to a very simple tautology like $p \rightarrow p$. So, it seems initially plausible that someone who is just encountering PL for the first time may not (yet) believe it to be true, despite believing the following obvious truth:

(OT) $\forall p(p \rightarrow p)$

The Foundations of Modality: From Propositions to Possible Worlds. Peter Fritz, Oxford University Press.
© Peter Fritz 2023. DOI: 10.1093/oso/9780192870025.003.0004

Thus, using b as a unary operator constant interpreted as expressing being believed by the relevant agent, one might consider the following a plausible claim:

$$b(\text{OT}) \wedge \neg b(\text{PL})$$

Using LL, the distinctness of these propositions follows:

$$\text{PL} \neq \text{OT}$$

One might be tempted to argue along these lines that propositions have to be individuated relatively finely. In particular, one might conclude that even sentences which are provable, and so equivalent in $\vdash$, may express distinct propositions.

This type of argument can be generalized very widely by iterating attitude reports, following an idea of Mates (1952). Let φ and ψ be distinct sentences. Let m be a unary operator expressing the property of being believed by a suitable agent, which can be taken to be a philosopher who is unsure about the proper treatment of attitude ascription. Let $>$ be a binary operator such that $p > q$ states that necessarily, everyone who believes p believes q. $\varphi > \varphi$ is clearly true, since necessarily, everyone who believes φ believes φ. Assume that this reasoning is available to the agent under consideration, whence $m(\varphi > \varphi)$. Consider now the claim $\varphi > \psi$, that necessarily everyone who believes φ also believes ψ. Depending on the choice of φ and ψ, it may be obvious that this is false. It may also not be obvious. But since φ and ψ are distinct sentences, it is at least conceivable that someone believes φ without believing ψ. As long as φ and ψ are distinct sentences, someone who is cautious about matters of attitude ascriptions therefore might refrain from believing that it is necessary that everyone who believes φ also believes ψ. Thus $m(\varphi > \psi)$ appears to be false. So:

$$m(\varphi > \varphi) \wedge \neg m(\varphi > \psi)$$

Using LL, one might use this line of argument to motivate the following schema:

(DISTINCTNESS) $\varphi \neq \psi$ (where φ and ψ are distinct sentences)

So, it might initially seem that arguments using attitude ascriptions are very informative, since they allow us to use plausible judgements to motive a strong principle like DISTINCTNESS. However, we will now see that DISTINCTNESS is *too strong*: it cannot possibly be true. Therefore, these arguments using attitude ascriptions overgenerate, and so must be based on some mistake. First, it is easy to see that DISTINCTNESS cannot be true in general. Consider any distinct propositional constants c and d. A statement like $c = d$ is true or false only relative to interpretations of c and d. No particular constraints are imposed on these interpretations. In particular, nothing rules out an interpretation on which c and d express the same proposition, such as $2 + 2 = 4$. On this interpretation, $c = d$ is true, contradicting DISTINCTNESS.

This point can be illustrated from another perspective. Consider the original case of Frege (1892b), involving Hesperus and Phosphorus. Intuitively, it seems that an agent can believe that Hesperus is Hesperus without believing that Hesperus is Phosphorus. In a language like $\mathcal{L}^*$ which provides singular terms, this can be used to give an argument analogous to the argument from attitude ascriptions considered above which concludes that Hesperus is not Phosphorus. But whatever one makes of this vexing case, it is a non-starter to accept the conclusion: there is no denying that Hesperus is Phosphorus. Frege cases cannot be used to establish the distinctness of individuals. On the intended interpretation of $\mathcal{L}$, formulas stand to propositions as individual terms stand to individuals. By analogy, there is therefore a good general reason to doubt that attitude ascriptions can be used to draw any straightforward conclusions about the individuation of propositions.

What exactly goes wrong with the argument motivating DISTINCTNESS depends on what exactly is going on with attitude ascriptions in English. This is not the place to attempt to come to any conclusions about this difficult matter. But it is worth illustrating briefly what happens to the argument for DISTINCTNESS on a couple of views of attitude ascriptions.

Consider first a view on which attitude contexts in English are sensitive to more than the proposition expressed by the complement clause; for an example, see Stalnaker (1999). On such a view, a sentence is associated not only with the proposition it expresses, but also a different value which one may call the *guise* in which it presents the proposition. What guises are may be left open at this point; examples are Fregean senses, or the sentences themselves. In this case, the stipulated logical principles of $\mathcal{L}^=$, in particular LL, rule out the possibility of interpreting an operator constant b of $\mathcal{L}^=$ in any way which would correspond to such an attitude context. There is therefore no way of settling the interpretation of b in such a way that our judgements about belief can be used to conclude anything about which propositions fall under b.

As a second example, consider a contextualist view, on which words like "believes" used in attitude ascriptions are massively context-sensitive; for an example, see Dorr (2014). On such a view, it may well be possible to interpret a constant b to correspond very closely to the English phrase "the agent believes that," given a determination of the context. But on these contextualist views, the usual Frege cases are accounted for using context shifts. Without substantial further commitments, one can therefore not use any intuitive judgements about the possibility of an agent believing φ without believing ψ to draw any conclusion to the effect that $\varphi \neq \psi$.

3.2 Structure and the Russell-Myhill Argument

We may not be able to rely unreflectively on all intuitive judgements about attitudes, but the reasons which we had for rejecting DISTINCT-NESS do not generalize to all cases in which attitude ascriptions motivate fine distinctions between propositions. To illustrate this, consider again the initial example of PL and OT: there is a salient difference between the sentences PL and OT which is not present in the case of arbitrary sentential constants c and d, namely, that PL and OT differ in their syntactic structure. If the syntactic structure of a sentence is reflected in the proposition it expresses, then the sentences PL and OT express different propositions, and the natural judgement that someone might believe OT without believing PL can be retained, even if some other natural judgements about (iterated) attitudes have to be rejected.

Such a structured conception of propositions is common in several areas of philosophy, although it is probably most often appealed to in the philosophy of language; see King (2019) for an overview. According to this kind of view, the proposition expressed by a sentence is a structured complex constituted by what the atomic terms of the sentence express. For example, the proposition that Jupiter is a planet is a structured complex constituted by the planet Jupiter and the property of being a planet. As a consequence, the proposition that Jupiter is a planet is distinct from the proposition that Mercury is a planet (since Jupiter is not Mercury), as well as being distinct from the proposition that Jupiter is a star (since to be a planet is not the same as to be a star).

This aspect of the structured view can be stated more generally in the formal language $\mathcal{L}$. Let $\varphi(\bar{x})$ be a formula with n free variables $\bar{x}$, and let $\bar{y}$ be a corresponding sequence of variables so that $\varphi(\bar{y})$ is a formula with n free variables $\bar{y}$. Then the proposition $\varphi(\bar{x})$ can only be the proposition $\varphi(\bar{y})$ if x_i is y_i, for all $i \leq n$. Since no identity connective for modalities is available in $\mathcal{L}^=$, this cannot be formalized in the intended strength here. But what is identical is functionally equivalent, so at least the following weaker principle can be formulated:

(STRUCTURE) $\varphi(\bar{x}) = \varphi(\bar{y}) \rightarrow x_i \equiv y_i$

To be clear, this principle also does not obviously capture the full structured view for other reasons; for example, it does not obviously tell us that $\neg p \neq (p \wedge p)$, which is plausibly part of the structured view. But we can set this issue aside for the moment, since we will now see that STRUCTURE is already problematic on its own: STRUCTURE can be shown to be false by a purely logical argument, as various instances of STRUCTURE are inconsistent in $\vdash^=$.

One example of an inconsistent instance of STRUCTURE states that if the proposition that m only applies to truths is the proposition that n only applies to truth, then m is (functionally equivalent to) n:

$$(\forall p(mp \rightarrow p) = \forall p(np \rightarrow p)) \rightarrow m \equiv n$$

The argument for the inconsistency goes back to Russell (1903, appendix B). It was rediscovered by Myhill (1958), so it is sometimes called the Russell-Myhill argument. Generalizing this argument, one can show that for every formula φ, it is provably false that for any properties of propositions m and n, if $\varphi(m)$ is $\varphi(n)$ then m is functionally equivalent to n:

Theorem 3.1. $\vdash^= \neg\forall m\forall n(\varphi(m) = \varphi(n) \rightarrow m \equiv n)$

Informally, the argument can be thought of as an application of Cantor's theorem. Just as any set has fewer elements than subsets, there are fewer propositions than unary properties of propositions. Consequently, there cannot be a unique proposition $\varphi(m)$ for every modality m. Of course, this informal line of argument does not suffice to establish Theorem 3.1. It is not clear how cardinality talk should be captured in the language $\mathcal{L}$, especially as $\vdash$ does not include any form of the axiom of choice. It is therefore important to produce an explicit derivation. This is relatively easily done using the following construction, by considering whether $r\varphi(r)$:

$$r := \lambda p.\neg\exists o((\varphi(o) = p) \wedge op)$$

However, merely presenting a derivation along these lines makes the argument seem somewhat like magic. To see how the definition of r can be arrived at, it is worth seeing how it follows by adapting Russell's argument for the inconsistency of naive set comprehension.

Let o be any modality (i.e., property of propositions), and assume that $\varphi(m)$ is a formula with a free operator variable m. According to the structured proposition view, o is the unique modality m such that $\varphi(o) = \varphi(m)$. Relative to the arbitrary choice of the formula φ, the proposition $\varphi(o)$ can therefore be understood as standing for o. In general, a proposition p stands for a modality o just in case p is $\varphi(o)$. Some propositions may not stand for any modality, but for every modality o, there is a unique proposition which stands for o, namely $\varphi(o)$, and $\varphi(o)$ stands for no other modality. For any propositions p and q, one can therefore state that there is a modality for which p stands and which applies to q, as follows:

$$q \text{ has } p := \exists o((\varphi(o) = p) \wedge oq)$$

With this, one can apply Russell's argument against naive set comprehension to properties. In particular, one can formulate the following operator:

$$r := \lambda p.\neg(p \text{ has } p)$$

Note that with the definition of "has" just given, this is exactly the same definition of r as above. r expresses the modality of being a proposition which does not fall under any modality it stands for. Russell's argument against naive set comprehension can now be adapted by considering the question whether the proposition $\varphi(r)$ standing for r falls under r.

If $r\varphi(r)$, then there is no modality o such that $\varphi(o) = \varphi(r)$ and $o\varphi(r)$. Since $\varphi(r) = \varphi(r)$, it follows that $\neg r\varphi(r)$, which contradicts the assumption. So $\neg r\varphi(r)$. It follows that there is a modality o such that $\varphi(o) = \varphi(r)$ and $o\varphi(r)$. Since o but not r applies to $\varphi(r)$, these modalities differ in their extension. Thus $\varphi(o) = \varphi(r)$ even though $o \not\equiv r$, which contradicts STRUCTURE.

Since this result is central in the overall argument of this book, it is worth giving a relatively explicit deduction (although some simple steps of truth-functional and quantificational reasoning are still omitted).

Proof of Theorem 3.1. By the following deduction in $\vdash^=$:

(1) $r\varphi(r) \rightarrow \neg\exists o((\varphi(o) = \varphi(r)) \wedge o\varphi(r))$ λC

(2) $r\varphi(r) \rightarrow \forall o((\varphi(o) = \varphi(r)) \rightarrow \neg o\varphi(r))$ 1, UE

(3) $r\varphi(r) \rightarrow ((\varphi(r) = \varphi(r)) \rightarrow \neg r\varphi(r))$ 2, UI

(4) $\neg r\varphi(r)$ 3, RI

(5) $\exists o((\varphi(o) = \varphi(r)) \wedge o\varphi(r))$ 4, λC

(6) $o\varphi(r) \rightarrow o \not\equiv r$ 4, UI

(7) $\exists o((\varphi(o) = \varphi(r)) \wedge o \not\equiv r)$ 5, 6

(8) $\neg\forall m\forall n(\varphi(m) = \varphi(n) \rightarrow m \equiv n)$ 7

□

The Russell-Myhill argument shows that propositions cannot be structured. This provides a first illustration of how the method of improving upon informal talk of propositions and modalities using $\mathcal{L}$ bears fruit. Using basic logical principles, non-obvious consequences can be derived. It would be a mistake to regard this as a sign that something has gone wrong. Recall that these basic principles are adopted stipulatively, as part of the metasemantics of $\mathcal{L}$. Assuming the introduction of $\mathcal{L}$ succeeds, the truth of these principles is therefore not in question. And the derivability of non-obvious consequences from these principles is a reason *for* thinking that the introduction succeeded, rather than *against* thinking so: such results show that the stipulative logical principles constrain the interpretation of the new language in a more substantial way than one might have thought. Thereby, they alleviate worries that the intended interpretation has not been completely settled. Such worries might be thought to be especially pressing if attitude contexts are set aside as anchors for the possible interpretations of operators, as argued by Lederman (forthcoming). Furthermore, confronted with questions which are not settled by the stipulative logical principles, it is often quite difficult to see how to find out what the correct answer is; this will be a major concern in the following. Results like the Russell-Myhill argument should therefore be cherished, rather than regarded with suspicion. In the present setting, there are therefore no reasons to consider ways of avoiding Theorem 3.1, such as ramifying the type theory or imposing predicativity restrictions on the quantificational logics; for examples of such responses to the Russell-Myhill argument, see Hodes (2015) and Walsh (2016), respectively.

Just as this section has explored retreating from DISTINCTNESS to STRUCTURE, given the inconsistency of DISTINCTNESS, one might consider restricting STRUCTURE to obtain a consistent but non-trivial principle which allows us to account for at least some central judgements about attitude ascriptions using the individuation of propositions. There are many options which could be explored. Surprisingly, many principles which appear to be weaker than STRUCTURE are likewise inconsistent. Two options along these lines are discussed in the following two supplementary sections.

3.3 Supplement: Logical Structure and Grounding

Since STRUCTURE is inconsistent, one might try to retreat to weaker forms of structured propositions. For example, one might note that the sentences PL and OT involve differences in their logical structure, in the sense that these sentences are built up from proposition letters using different applications of logical connectives. One might therefore consider a view on which propositions have at least *logical structure*, which entails that, e.g., negations $\neg p$ and $\neg q$ are identical only if the negated propositions p and q are identical, even if we cannot conclude for an arbitrary modality m that mp and mq are identical only if p and q are identical.

However, even this logical structure view is inconsistent, as shown in (Fritz 2023). The inconsistency can be established using any binary sentential connective $\circ$ and any propositional quantifier Qp. For concreteness, we consider conjunction ($\wedge$) and the universal quantifier ($\forall p$) in the following. In both cases, we need to formulate what it takes for propositions to inherit the logical structure which the relevant connective contributes to sentences. In the case of conjunction, the principle follows the case of negation just mentioned: roughly, conjunctions are identical only if their conjuncts are identical. Since conjunction is binary, a choice point arises which does not arise in the unary case of negation. We can require either that conjunctions are identical only if their first conjuncts are identical and their second conjuncts are identical, or that a first conjunction is identical to a second conjunction only if each conjunct of the first is a conjunct of the second, and vice versa. The two options differ in whether the order of conjuncts is part of the logical structure of conjunctions. Here, it does not matter which version is adopted, since the first version is stronger than the second version, and the second version suffices for the inconsistency. We therefore use the second version. Formally, it can be stated as follows:

(LS$\wedge$) $(p \wedge q) = (r \wedge s) \rightarrow (p = r \wedge q = s) \vee (p = s \wedge q = r)$

Turning to propositional quantifiers, there are also two natural options. (There is a third option if quantifiers are treated as higher-order predicates; see Fritz (2020) and Goodman (2023). Since $\mathcal{L}$ includes quantifiers as variable binders, it does not arise here.) One option is to say that the quantified propositions $\forall p \varphi$ and $\forall p \psi$ are identical only if the properties of satisfying the two complement clauses are identical. In $\mathcal{L}$, we can capture at least a substantial component of this by saying that $\forall p \varphi$ is $\forall p \psi$ only if $\lambda p. \varphi$ and $\lambda p. \psi$ are functionally equivalent. However, it is easy to see that this is inconsistent by itself, by a minor variation of Theorem 3.1. More carefully, one might think of a universal quantification $\forall p\varphi$ as a long conjunction, containing as conjuncts just the instances of $\forall p\varphi$. These instances are the propositions $\varphi(q)$ for every proposition q. Corresponding to LS$\wedge$, one might therefore propose that quantified propositions $\forall p\varphi$ and $\forall p\psi$ are identical only if their instances are the same, disregarding any order in which they may be put. So, $\forall p\varphi$ is $\forall p\psi$ only if for every p, there is a q such that $\varphi(p)$ is $\psi(q)$, and vice versa. By the symmetry of identity, we can omit the vice versa clause, and so formulate the principle as follows:

(LS$\forall$) $(\forall p\varphi(p) = \forall p\psi(p)) \rightarrow \forall p\exists q(\varphi(p) = \psi(q))$

The schematic principle LS$\forall$ is not inconsistent on its own, and neither is LS$\wedge$; see Fritz (2023) and Goodman (2023, appendix D). However, the two principles are jointly inconsistent. This can be derived as a consequence of Theorem 3.1. Since the proof is relatively straightforward, it is worth reproducing it here. The first step is to introduce the following abbreviation:

$$\varphi \hat{\wedge} \psi := ((\varphi \wedge \neg\varphi) \wedge (\varphi \wedge \varphi)) \wedge (\psi \wedge \psi)$$

It can now be shown that from $\varphi \hat{\wedge} \psi$, φ and ψ can be recovered *in order*:

Lemma 3.2. LS$\wedge \vdash^= (\varphi \hat{\wedge} \varphi') = (\psi \hat{\wedge} \psi') \rightarrow (\varphi = \psi \wedge \varphi' = \psi')$.

Proof. Assume $(\varphi \hat{\wedge} \varphi') = (\psi \hat{\wedge} \psi')$, i.e.:

(1) $(((\varphi\wedge\neg\varphi)\wedge(\varphi\wedge\varphi))\wedge(\varphi'\wedge\varphi')) = (((\psi\wedge\neg\psi)\wedge(\psi\wedge\psi))\wedge(\psi'\wedge\psi'))$

Using LS$\wedge$, we can reason as follows. First, as $\varphi \neq \neg\varphi$, also $(\varphi \wedge \neg\varphi) \neq (\varphi \wedge \varphi)$. So:

(2) $((\varphi \wedge \neg\varphi) \wedge (\varphi \wedge \varphi)) \neq (\psi' \wedge \psi')$

From (1) and (2), the following two claims can be inferred:

(3) $((\varphi \wedge \neg\varphi) \wedge (\varphi \wedge \varphi)) = ((\psi \wedge \neg\psi) \wedge (\psi \wedge \psi))$
(4) $(\varphi' \wedge \varphi') = (\psi' \wedge \psi')$

By (3), since $(\varphi \wedge \neg\varphi) \neq (\psi \wedge \psi)$, $(\varphi \wedge \varphi) = (\psi \wedge \psi)$, whence $\varphi = \psi$. By (4), $\varphi' = \psi'$.   □

Using this lemma, LS∧ and LS∀ can be seen to entail that from $\forall p(mp \wedge p)$, the propositions falling under m can be recovered, and by the Russell-Myhill argument, this is inconsistent:

Lemma 3.3. LS∧, LS∀ $\vdash^= \forall p(mp \wedge p) = \forall p(np \wedge p) \rightarrow m \equiv n.$

Proof. Assume $\forall p(mp \wedge p) = \forall p(np \wedge p)$. By LS∀, it follows that:

$$\forall p \exists q((mp \wedge p) = (nq \wedge q))$$

So by Lemma 3.2:

$$\forall p \exists q(mp = nq \wedge p = q)$$

Thus $\forall p(mp = np)$, whence $m \equiv n$.   □

Proposition 3.4. LS∧ *and* LS∀ *are jointly inconsistent in* $\vdash^=$.

Proof. Immediate by Lemma 3.3 and Theorem 3.1.   □

The stipulative logical principles of $\mathcal{L}^=$ therefore not only rule out STRUCTURE, but even apparently weaker versions of structured proposition views, including the view that propositions reflect the logical structure of Boolean connectives and propositional quantifiers. Furthermore, Proposition 3.4 entails that there cannot be both a relation C of being a conjunct and a relation I of being an instance of a universal quantification satisfying the following intuitively plausible principles:

(CONJUNCT) $(Cp(q \wedge r)) \leftrightarrow (p = q) \vee (p = r)$
(∀-INSTANCE) $(Ip\forall q\varphi(q)) \leftrightarrow \exists q(p = \varphi(q))$

CONJUNCT entails LS∧. Informally, if $p \wedge q$ is $r \wedge s$, then any proposition t is a conjunct of the former iff it is a conjunct of the latter. So by CONJUNCT, any t is one of p and q iff t is one of r and s, which means that either p is r and q is s, or p is s and q is r. This argument is easily turned into a formal derivation in $\vdash^=$. Similarly, ∀-INSTANCE entails LS∀. So by Proposition 3.4, CONJUNCT and ∀-INSTANCE are jointly inconsistent. Yet, as noted in (Fritz 2023), each of CONJUNCT and ∀-INSTANCE is consistent on its own. These observations can be used to draw a couple of further consequences.

The first consequence concerns the metasemantics of languages like $\mathcal{L}^*$ and $\mathcal{L}$. As mentioned briefly above, Lederman (forthcoming) discusses the worry that the metasemantic stipulations for such languages might be insufficient if too many English sentential contexts are excluded from serving a constraining role on the metasemantics of operator expressions and quantifiers binding variables in operator position. One might be especially worried if little is said about exactly which such contexts do serve this role. In response to this, one might hope to offer at least a rough criterion of logical consistency: an English sentential context is taken to constrain the metasemantics of operator expressions just in case central intuitive judgements concerning the context do not contradict the stipulative logical principle of the relevant language, such as those encoded in $\vdash^=$ in the case of $\mathcal{L}^=$. Many contexts which ascribe propositional attitudes such as belief are excluded by this criterion, for the reasons discussed earlier in this chapter. Truth-functional contexts are included, as are arguably some modal contexts.

Admittedly, this criterion is somewhat vague, since it is not always clear what counts as a central intuitive judgement. But the criterion is clear enough for us to see that it is unworkable; this is what the cases of "conjunct" and "instance" show. (To be clear, these English words are not sentential contexts. Yet neither is "necessary" or "belief," which are clearly in the intended range of the connection. There is no need for us to go into the syntactic details of, e.g., the status of that-clauses as complements, since the metasemantic constraints of natural language correlates are not meant to be guided by any hard and fast syntactic correspondences.) Plausibly, the proposed criterion rules "conjunct" and "instance of a universal quantification" as included in the constraining role, since the central intuitive judgements are captured by CONJUNCT and ∀-INSTANCE, which we have noted to be consistent. Yet, they cannot both be included, since the inconsistency of CONJUNCT and ∀-INSTANCE shows that $\mathcal{L}^=$ cannot contain both constants C and I satisfying these two principles.

The second consequence concerns the ideology of metaphysical grounding, discussed in the Introduction. The inconsistency of CONJUNCT and ∀-INSTANCE shows that there cannot be relations C and I satisfying these principles, respectively. *A fortiori*, there cannot be a single relation $\prec$ satisfying both of them. Formally, the variants of

CONJUNCT and ∀-INSTANCE for a single relation ≺ are therefore jointly inconsistent as well:

$$(\wedge\prec) \quad (p \prec (q \wedge r)) \leftrightarrow (p = q) \vee (p = r)$$
$$(\forall\prec) \quad (p \prec \forall q\varphi(q)) \leftrightarrow \exists q(p = \varphi(q))$$

Yet, these are the natural principles of immediate non-factive partial grounding of conjunction and universal quantification suggested by the relevant discussions in the grounding literature, such as Fine (2012): conjunctions are grounded in just their conjuncts, and universal quantifications are grounded in just their instances. A variant of this argument can also be given for a factive relation of ground, using principles restricted to true propositions, by adding an analog of $\wedge\prec$ for disjunction. Similarly, the argument can be refined to assume only a slightly weaker form of $\forall\prec$. The details can be found in (Fritz 2022).

What does this tell us about grounding? The discussion of grounding in metaphysics has sometimes been taken to show that propositions have to be individuated relatively finely. For example, $\top \wedge \bot$ and $\bot \wedge \bot$ are both provably false. According to $\wedge\prec$, they have to be distinct. According to this principle, $\top$ is a ground of $\top \wedge \bot$, and $\bot$ is the only ground of $\bot \wedge \bot$. $\top$ cannot be identical to $\bot$, since the former is true and the latter is false. Therefore, if $\wedge\prec$ is true, there are provably equivalent sentences which express distinct propositions. The argument for the inconsistency of $\wedge\prec$ and $\forall\prec$ can be seen as pushing this kind of observation further: we can think of it as showing that $\wedge\prec$ and $\forall\prec$ require propositions to be so finely individuated as to run into the same kind of inconsistency as the logical structure principles. We might sum this observation up as showing that central grounding principles require propositions to be finely individuated to an inconsistent degree.

In the context of $\mathcal{L}^=$, since there is no relation satisfying paradigmatic principles of grounding, it is plausible that metasemantic stipulations corresponding to those made by grounding theorists fail to single out any particular relation. In this setting, the putative connective of grounding is, like the English phrases used to introduce it such as "explains," "makes true," and "in virtue of," plausibly sensitive to representational features of sentences, perhaps similar to contexts attributing propositional attitudes like belief, despite the intentions of grounding theorists to the contrary.

This conclusion illustrates how identity-first metaphysics, outlined in the Introduction, can be used to make progress on difficult questions in metaphysics. Little progress can be expected from using our intuitions to argue for and against the view that grounding is a relation of metaphysics, as opposed to tracking features of our representation of reality. By focusing on the individuation of propositions, and the demands of grounding on such individuation, it is possible to give a general argument that grounding must be of the latter, representational, nature, and so should not be added to our metaphysical vocabulary. This in turn has consequences for our views about propositional individuation, since it removes one further potential reason for thinking that propositions are individuated relatively finely.

3.4 Supplement: Closed Structure

Above, we saw how Mates's argument leads to the DISTINCTNESS schema. This led to a discussion of the popular view that propositions are structured, partly captured by STRUCTURE. We noted that there is an important sense in which DISTINCTNESS goes beyond STRUCTURE, as it implausibly rules out propositional constants expressing the same proposition. But there is also a sense in which STRUCTURE goes beyond DISTINCTNESS. This is because STRUCTURE involves free variables, implicitly bound by universal quantifiers. It is in fact not so clear that Mates's argument can be used to motivate this quantified principle, since Mates's argument as formulated above is schematic with respect to *sentences*, i.e., formulas without free variables. This poses the question whether the inconsistency of the schematic principle STRUCTURE can be avoided by restricting it to closed instances, i.e., instances without free variables.

Recall that the Russell-Myhill argument shows the following instance of STRUCTURE to be inconsistent:

$$(\forall p(mp \to p) = \forall p(np \to p)) \to m \equiv n$$

For any m and n which are not co-extensive, one might try to argue, along the lines of Mates's argument, that someone may believe that necessarily everyone who believes $\forall p(mp \to p)$ also believes $\forall p(mp \to p)$, without that person also believing that necessarily everyone who believes $\forall p(mp \to p)$ believes $\forall p(np \to p)$. But in this case, the matter is much less straightforward, since the relevant attitudes are *de re* with respect to arbitrary modalities m and n. It is far from clear that for every modality, it is possible to have the relevant attitudes. It is at least worth entertaining the possibility that some modalities are ineffable, in the sense that the relevant propositions involving them cannot be entertained.

One might therefore consider a schematic principle of CLOSED STRUCTURE, which restricts STRUCTURE to closed instances. To illustrate this, since m and n are free variables in the instance of STRUCTURE displayed above, this formula is not an instance of CLOSED STRUCTURE. However, an instance of STRUCTURE like $(\Box T = \Box\Box T) \to (T = \Box T)$ is also an instance of CLOSED STRUCTURE, since it has no free variables.

In the derivation of an inconsistency from STRUCTURE, the Russell-Myhill argument explicitly identifies a modality r, but only proves—without explicitly specifying a witness—the existence of another modality o not coextensive with r such that $\varphi(r)$ is $\varphi(o)$. This raises the question whether CLOSED STRUCTURE is consistent. In Fritz et al. (2021), it is shown that such a principle is indeed consistent in a language along the lines of $\mathcal{L}^*$ (and therefore also its fragment $\mathcal{L}$). However, it is there also shown that the inconsistency arises again if the principle is formulated in a language enriched by the resources to speak plurally about propositions. An extension along these lines will play an important role in Chapter 7. For now, it suffices merely to note that this route to a restricted form of STRUCTURE is therefore also untenable.

This conclusion adds further support to the conclusion reached above. On the present conception, propositions simply are not structured, and it is a mistake to think that attitude ascriptions provide straightforward reasons to distinguish them very finely.

4

Classicism

The last chapter discussed how the Russell-Myhill argument shows several natural theories of structured propositions to be inconsistent. Among theories of propositional identification, such theories are on one end of a spectrum that goes from coarse to fine. In exploring metaphysical views, it is useful to focus—at least initially—on simple and strong views. Such views are naturally expected to lie at the ends of this spectrum. With various theories on the fine-grained end of the spectrum having been showed to be problematic, it is therefore natural to consider simple and strong coarse-grained views. This chapter singles out one such view.

4.1 Extensionalism

There is one coarse-grained view which is very simple and very strong, which we will call EXTENSIONALISM. According to this principle, materially equivalent propositions are identical. Thus, according to this view, there are only two propositions. We might think of this view as identifying propositions with truth-values. It can be stated formally as the following principle, which Suszko (1975) calls the *Fregean Axiom*:

(EXTENSIONALISM) $(p \leftrightarrow q) \rightarrow (p = q)$

In fact, EXTENSIONALISM is the coarsest individuation of propositions available: it is straightforward to prove in $\vdash^=$ that there are distinct propositions:

$$\exists p \exists q (p \neq q)$$

The Foundations of Modality: From Propositions to Possible Worlds. Peter Fritz, Oxford University Press.
© Peter Fritz 2023. DOI: 10.1093/oso/9780192870025.003.0005

This follows from the existence of a true proposition, such as $\top$, and a false proposition, such as $\bot$. PROPOSITIONAL NIHILISM ($\neg\exists p(p = p)$) and PROPOSITIONAL MONISM ($p = q$) are therefore inconsistent, and so false. Any exploration of coarse-grained views must therefore start with EXTENSIONALISM.

Yet, EXTENSIONALISM faces some serious difficulties. A natural objection notes that some truths are necessary while others are contingent. For example, it is usually taken to be necessary that $2 + 2 = 4$, but contingent that Mars has two moons. From this, one concludes that there are at least two truths, and so that material equivalence does not suffice for propositional identity. The natural way of turning this into an argument against EXTENSIONALISM takes a unary operator constant $\square$, interpreting it as necessity, and two propositional constants p and q, interpreting them as the proposition that $2 + 2 = 4$ and the proposition that Mars has two moons, respectively. Then

$$p \wedge q \wedge \square p \wedge \neg\square q$$

appears to be true, from which the failure of EXTENSIONALISM can be deduced in $\vdash^=$.

While this argument is compelling, there is a possible way of resisting it. What is expressed by modal terms in English depends greatly on context. Furthermore, in many contexts, modal terms express epistemic statuses: to say that something is necessary, or must be the case, amounts to saying that it is known by some salient agent or agents. As argued in the previous chapter, such attitude ascriptions cannot directly be used to argue for propositional distinctness: they may lead to consequences about propositional distinctness which are inconsistent in $\vdash^=$, which means that the operator constants of $\mathcal{L}$ cannot be interpreted in a way which captures the relevant English attitude ascriptions. The same type of argument might rule out any interpretation of $\square$ as a notion of necessity which distinguishes $2 + 2$ being 4 from Mars having two moons.

However, as Kripke (1980 [1972]) argued, there are plausibly contexts in which modal terms are used not in an epistemic sense, but in a worldly or metaphysical sense. The modalities expressed in these contexts are concerned with ways the world could be, in a sense which has nothing to do with anyone's knowledge, belief, or other attitude. Taking a term from the linguistic literature on modals, let me call such readings of modal terms *circumstantial*. (There is no agreement in this literature on how to classify the different readings of modal terms, as discussed by Portner (2009, §4.1). The usage of "circumstantial" adopted here consequently deviates from certain uses of this term in the literature.) Importantly, the claim that there are such circumstantial modal notions is weaker than the claim that there is a single notion of metaphysical necessity as widely appealed to in contemporary metaphysics, a claim with which the next chapter will be concerned. To say that there are circumstantial modalities does not require there to be any one such modality which is distinguished in a special way. Consider, then, one such use of modal terms in English. Plausibly, there is such a sense which counts $2 + 2 = 4$ as being necessary without counting Mars having two moons a necessary, for example a notion of physical necessity. Again, it does not matter for present purposes if this singles out a particular modality—what matters is only that there is some such modality. In contrast to the case of attitude ascriptions, there are therefore strong reasons to think that there is a nontrivial modal interpretation of $\Box$, and so that EXTENSIONALISM is false.

Some might want to resist even the weak assumption that there is *some* circumstantial modal notion which can be used to interpret an operator like $\Box$ to show that EXTENSIONALISM is false. A supplementary section 4.6 considers and responds to two objections along these lines.

The argument against EXTENSIONALISM given above generalizes to an argument showing that there are not only more than two, but infinitely many propositions. For example, it is plausible that for every natural number n, it is possible that there are exactly n elementary particles, such as electrons. Let p_n be the proposition that there are exactly n elementary particles. Then for any distinct natural numbers $n \neq m$, it is not possible for there to be both exactly n elementary particles and exactly m elementary particles. Using $\Diamond$ for possibility, we therefore have:

$$\neg\Diamond(p_n \wedge p_m)$$

Assume for contradiction that $p_n = p_m$. Then with Leibniz's Law, we obtain the following conclusion:

$$\neg\Diamond(p_n \wedge p_n)$$

But this contradicts that for any natural number n, it is possible for there to be exactly n elementary particles. Therefore, there is a distinct proposition p_n for every natural number n, and so infinitely many propositions. To express this in $\mathcal{L}^{=}$, we first state, for any given natural number n, that there are at least n propositions, as follows:

$$(\mathrm{E}_n) \quad \exists p_1 \ldots \exists p_n \bigwedge_{1 \leq i < j \leq n} (p_i \neq p_j)$$

Here, $\bigwedge_{1 \leq i < j \leq n} (p_i \neq p_j)$ indicates a conjunction of formulas of the form $p_i \neq p_j$, for parameters i and j such that $1 \leq i < j \leq n$. That there are infinitely many propositions can then be captured using the schematic principle which has as its instances E_n, for every natural number n. We can think of this as the set of these instances, i.e.:

$$\infty := \{\mathrm{E}_n : n \in \mathbb{N}\}.$$

4.2 Boolean Connectives

With the focus on coarse-grained theories which are especially strong and simple, the falsity of EXTENSIONALISM poses the question whether there is any way of singling out another coarse-grained theory which is simple and strong, while individuating propositions finely enough to allow for the modal distinctions with which EXTENSIONALISM conflicts. The remainder of this chapter will develop such a theory. Since the goal is to find a distinguished coarse-grained theory, it is natural to focus on theories which can be stated in terms of identifications: in this case, strength can be understood as deductive strength, and stronger theories will individuate propositions more coarsely. To arrive at simple theories, one may then focus on simple identifications. The following are three paradigmatic principles of this kind:

$$p = \neg\neg p$$
$$(p \wedge q) = (q \wedge p)$$
$$p = (p \vee p)$$

Taking these to be implicitly about all propositions p and q, the first states that any proposition is the negation of its negation, the second that conjunctions are insensitive to the order of the conjuncts, and the third that propositions are invariant under being disjoined with themselves.

Principles such as these three are natural first cases to consider when developing general views of propositional identity. They therefore serve as paradigms of simple principles governing propositional identity. There is also a natural pattern into which they all fall: they have the form of an equation $\varphi = \psi$ in which φ and ψ are formulas involving only propositional variables and Boolean connectives. It will be useful to introduce a general way of referring to such restricted classes of equations:

Definition 4.1. *Let X be a language. An X-equation is a formula of the form $\varphi = \psi$, with φ and ψ formulas of X. An X-equational theory is a set of X-equations.*

To isolate the relevant equations, we define the relevant language:

Definition 4.2. *Let $\mathcal{L}^B$ be the fragment of $\mathcal{L}$ consisting of the formulas containing only propositional variables and Boolean connectives.*

Thus, a first attempt at operationalizing the criterion of simplicity is to consider only sets of $\mathcal{L}^B$-equations, i.e., $\mathcal{L}^B$-equational theories. Strength can be operationalized simply in terms of deductive strength: Γ is at least as strong as Δ if every member of Δ can be derived from Γ. One can now ask a purely formal question: what is the structure of $\mathcal{L}^B$-equational theories, ordered by deductive strength?

It turns out that there is a unique strongest consistent $\mathcal{L}^B$-equational theory. Thus, simplicity and strength as operationalized here, together with consistency, serve to distinguish a unique candidate theory: among the consistent theories satisfying the simplicity constraint, there is a single theory which is strongest, and indeed includes all the other theories. It consists of the three equations stated above, as well as any other $\mathcal{L}^B$-equation $\varphi = \psi$ such that φ and ψ are equivalent according to $\vdash^=$ (which, for $\mathcal{L}^B$-equations, is easily seen to be the same as being equivalent according to classical propositional logic).

Proposition 4.3. *An $\mathcal{L}^B$-equational theory Γ is consistent in $\vdash^=$ if and only if $\vdash^= \varphi \leftrightarrow \psi$ for every equation $\varphi = \psi$ in Γ.*

Proof. For the right to left direction, assume $\vdash^= \varphi \leftrightarrow \psi$ for every equation $\varphi = \psi$ in Γ, and consider any standard possible worlds model $\mathfrak{M}$. By an induction on the length of derivations, it follows from Lemmas 2.10 and 2.12 that every theorem of $\vdash^=$ is valid on $\mathfrak{M}$. If $\varphi = \psi$ is a member of Γ, then $\varphi \leftrightarrow \psi$ is provable in $\vdash^=$, and so valid on $\mathfrak{M}$; by simplicity, it follows that $\varphi = \psi$ is valid on $\mathfrak{M}$ as well. The consistency of Γ in $\vdash^=$ therefore follows by Lemma 2.8.

For the left to right direction, consider any $\mathcal{L}^B$-equation $\varphi = \psi$ such that $\nvdash^= \varphi \leftrightarrow \psi$. It suffices to show that $\varphi = \psi$ is inconsistent. Since $\varphi \leftrightarrow \psi$ is not provable in $\vdash^=$, it is no tautology. Thus there is a truth-value assignment on which $\varphi \leftrightarrow \psi$ comes out false. There is then a corresponding substitution σ which maps every propositional variable to $\top$ or $\bot$ such that the negation of $\sigma(\varphi) \leftrightarrow \sigma(\psi)$ is a tautology. (A substitution is a mapping σ on formulas which respects the logical connectives, so that, e.g., $\sigma(\neg\varphi) = \neg\sigma(\varphi)$.) But this biconditional follows from $\varphi = \psi$, so $\varphi = \psi$ is inconsistent. □

It follows from this that standard possible worlds models validate every consistent $\mathcal{L}^B$-equational theory. The theory consisting of $\varphi = \psi$ for $\vdash^= \varphi \leftrightarrow \psi$ is the unique strongest such theory. As such, it is uniquely distinguished among $\mathcal{L}^B$-equational theories according to the present criteria of theory choice. It will therefore be worth introducing a name for it:

(PROPOSITIONAL BOOLEANISM)
$\phi = \psi$ (where $\varphi, \psi \in \mathcal{L}^B$ and $\vdash^= \varphi \leftrightarrow \psi$)

The remainder of this chapter will consider ways of extending the argument for this equational theory to equational theories in more inclusive languages.

It is worth noting that PROPOSITIONAL BOOLEANISM has been isolated before, in particular by Suszko (1975, see pp. 198–200 and en. XXXIX) under the label WB. A closely related view is also discussed by Dorr (2016) under the label *Booleanism*. PROPOSITIONAL BOOLEANISM is not only consistent on its own, but consistent with the existence of infinitely many propositions, as captured by ∞; this can be shown using possible worlds models based on infinite sets of worlds. PROPOSITIONAL BOOLEANISM on its own is consistent with EXTENSIONALISM, but of course no longer so once ∞ is added. These observations on consistency apply equally to the stronger equational theories to which we now turn.

4.3 Quantifiers

The argument in the last section implements the requirement of simplicity by considering only $\mathcal{L}^B$-equational theories. Roughly speaking, this means that the argument only considers principles concerning the interaction of Boolean operators with identity. But analogous to the three paradigmatic equations considered above, one might also be interested in equations such as the following:

$$\forall p\, p = \forall q\, q$$

$\mathcal{L}^B$-equational theories are effectively silent on such questions. It is therefore natural to ask whether the argument can be adapted to a theory which covers such equations as well.

There is an obvious way of pursuing such an extension, which is simply by admitting equations of the kind just displayed in the theories under consideration. Starting with only propositional quantifiers, the relevant equations can be isolated using the following fragment of $\mathcal{L}$:

Definition 4.4. *Let $\mathcal{L}^{BQ}$ be the fragment of $\mathcal{L}$ consisting of the formulas containing only propositional variables, Boolean connectives, and propositional quantifiers.*

Considering $\mathcal{L}^{BQ}$-equational theories amounts to adopting a slightly less restrictive criterion of simplicity. Comparing theories still in terms of deductive strength, this raises the question whether there is still a unique strongest consistent theory among $\mathcal{L}^{BQ}$-equational theories. This turns out to be the case, since propositional quantifiers are eliminable in ⊢, as will now be shown. The basic idea behind this elimination is that in

$\mathcal{L}^{BQ}$, materially equivalent propositions cannot be distinguished. Thus, φ holds for all p just in case it holds for both $\top$ and $\bot$. $\forall p\varphi$ is therefore equivalent to $\varphi[\top/p] \wedge \varphi[\bot/p]$. The following definition and results make this idea precise, and show how to extend Proposition 4.3 to $\mathcal{L}^{BQ}$:

Lemma 4.5. $\vdash (\psi \leftrightarrow \chi) \rightarrow (\varphi(\psi) \leftrightarrow \varphi(\chi))$, *for all formulas* φ, ψ, χ *of* $\mathcal{L}^{BQ}$.

Proof. By induction on the complexity of φ. □

Definition 4.6. *Let* $\cdot^B : \mathcal{L}^{BQ} \rightarrow \mathcal{L}^B$ *be the recursive mapping with the only non-trivial conditions:*

$$(\forall p\varphi)^B := \varphi^B[\top/p] \wedge \varphi^B[\bot/p]$$
$$(\exists p\varphi)^B := \varphi^B[\top/p] \wedge \varphi^B[\bot/p]$$

Proposition 4.7. $\vdash \varphi \leftrightarrow \varphi^B$, *for every formula* φ *of* $\mathcal{L}^{BQ}$.

Proof. By induction on the complexity of φ. Only the cases of quantifiers are interesting. Consider exemplarily the case of $\forall p$. We need to show the following:

$$\vdash \forall p\varphi \leftrightarrow \varphi^B[\top/p] \wedge \varphi^B[\bot/p]$$

The left to right direction follows by UI and the induction hypothesis. For the right to left direction, note that by Lemma 4.5, we have:

$$\vdash (p \leftrightarrow \top) \rightarrow (\varphi[\top/p] \rightarrow \varphi)$$
$$\vdash (p \leftrightarrow \bot) \rightarrow (\varphi[\bot/p] \rightarrow \varphi)$$

Since $\vdash (p \leftrightarrow \top) \vee (p \leftrightarrow \bot)$, we can conclude from this:

$$\varphi[\top/p] \wedge \varphi[\bot/p] \rightarrow \varphi$$

As p is not free in the antecedent, standard quantificational reasoning allows us to prefix the consequent with $\forall p$. The right to left direction therefore follows by induction hypothesis. □

Proposition 4.8. *An* $\mathcal{L}^{BQ}$*-equational theory* Γ *is consistent in* $\vdash^=$ *if and only if* $\vdash^= \varphi \leftrightarrow \psi$ *for every equation* $\varphi = \psi$ *in* Γ.

Proof. The right to left direction can again be shown using standard possible worlds models. For the left to right direction, consider an $\mathcal{L}^{BQ}$-equation $\varphi = \psi$ such that $\nvdash^= \varphi \leftrightarrow \psi$. By Proposition 4.7, it follows that $\nvdash \varphi^B \leftrightarrow \psi^B$. So as in the proof of Proposition 4.3, there is a substitution σ such that $\vdash \neg\sigma(\varphi^B \leftrightarrow \psi^B)$. But as $\varphi = \psi \in \Gamma$, $\varphi \leftrightarrow \psi$ follows from Γ, whence $\varphi^B \leftrightarrow \psi^B$ and so $\sigma(\varphi^B \leftrightarrow \psi^B)$ do as well. So Γ is inconsistent.   □

Thus, by eliminating quantifiers, the argument from the previous section can be extended to propositional quantifiers. The set of $\mathcal{L}^{BQ}$-equations $\varphi = \psi$ such that $\vdash^= \varphi \leftrightarrow \psi$ is the unique strongest consistent $\mathcal{L}^{BQ}$-equational theory, and valid on standard possible worlds models.

4.4 The Rule of Equivalence

The theories discussed in the last two sections are naturally seen as fragments of a more comprehensive theory, consisting of equations corresponding to provable biconditionals in the full language $\mathcal{L}$. Call this the EQUIVALENCE SCHEMA, in short ES:

(ES) $\varphi = \psi$ (where φ, ψ formulas of $\mathcal{L}^=$ such that $\vdash^= \varphi \leftrightarrow \psi$)

The above results raise the question whether an analogous result can be proven for ES. What is easily seen is that the schema is consistent with ∞, and so does not entail EXTENSIONALISM, since ES is valid on standard possible world models. But it cannot be shown that an $\mathcal{L}^=$-equational theory is consistent in $\vdash^=$ just in case it consists only of equations corresponding to provable biconditionals. Moreover, there is no single deductively strongest consistent $\mathcal{L}^=$-equational theory. Indeed, there are two formulas of $\mathcal{L}^=$ which involve identity as the only logical connective, and which are individually consistent and jointly inconsistent. This is shown in the supplementary section 4.7.

Despite these limitations, ES is a highly natural extension of the earlier two theories, and likewise simple and strong. Like the sub-theory containing only $\mathcal{L}^B$-equations, an analogous theory in a setting which omits quantifiers was already considered by Suszko (1975, see p. 200 and en. XXXIX), under the label of WT.

Although ES is quite strong, there is one more natural strengthening. According to ES, every biconditional provable in $\vdash^=$ determines a true equation. One may then naturally ask about biconditionals which can be proven in $\vdash^=$ from ES. The supplementary section 4.8 shows that not every such biconditional determines an equation which can itself be proven in $\vdash^=$ from ES. For instance, although $\forall p(\exists q(q = p) = \top) \leftrightarrow \top$ can be proven in $\vdash^=$ from ES, the equation $\forall p(\exists q(q = p) = \top) = \top$ cannot so be proven.

The schema ES can therefore be strengthened iteratively, by strengthening $\vdash^=$ in such a way that every equation corresponding to a provable biconditional is provable as well. To state this more rigorously, define the following rule of proof, called the RULE OF EQUIVALENCE, in short RE:

(RE) $\varphi \leftrightarrow \psi / \varphi = \psi$

This is a natural rule, which was already adopted by Cresswell (1965), and more recently by Bacon (2018a, p. 746).

Let CLASSICISM be the theory consisting of the $\mathcal{L}^=$-formulas derivable in $\vdash^= + $ RE, the proof system obtained by adding the RULE OF EQUIVALENCE to the axioms and rules of $\vdash^=$. Again, CLASSICISM is valid on standard possible worlds models, and so consistent with ∞. The label CLASSICISM is taken from Bacon and Dorr (forthcoming), who define a corresponding theory in a richer language. Due to the difference in language, their theory is not exactly the same as the present theory. It would therefore be appropriate to label the present view $\mathcal{L}^=$-CLASSICISM. But since this will be the only version of CLASSICISM considered here, the qualification will be left tacit.

CLASSICISM is the natural endpoint of a sequence of stronger and stronger equational theories in $\mathcal{L}^=$ which encode strong and simple coarse-grained views. As it is valid on standard possible worlds models, it is compatible with a metaphysically modal interpretation of $\Box$ of $\mathcal{L}$; in particular, it does not entail EXTENSIONALISM. CLASSICISM is therefore a good starting point for the development of a coarse-grained theory of propositional individuation.

Such a development of a coarse-grained theory will be carried out in the following chapters. Before delving into it, it is important to stress the difference between the theory CLASSICISM and the rule RE. The present discussion has motivated CLASSICISM as an appealing theory of propositional individuation, and therefore provided some grounds for the truth of its members. CLASSICISM is defined as the theorems provable in a certain proof system $\vdash^= +$ RE, just like the theorems of $\vdash^=$. But the argument for the truth of the members of CLASSICISM is of a very different character from the argument for the truth of the theorems of $\vdash^=$. In the latter case, it was argued that all axioms are true (under every interpretation of the constants and variables), and that the rules of proof preserve this status. No argument has been given for the claim that RE preserves being true under every interpretation of the constants and variables, nor for the claim that it preserves being logically true on any special sense of logical truth. Indeed, additional resources will later be introduced which satisfy principles of a general logical character such that closing them under RE would lead to EXTENSIONALISM. RE should therefore be considered to play a merely instrumental role in specifying CLASSICISM.

4.5 A Preliminary Assessment

CLASSICISM individuates propositions relatively coarsely: according to this view, any two sentences whose material equivalence can be proved from CLASSICISM express the same proposition. This means in particular that $\top$ is the unique proposition expressed by every sentence provable from CLASSICISM, and $\bot$ is the unique proposition expressed by every sentence whose negation is provable from CLASSICISM.

Another immediate and noteworthy consequence of CLASSICISM arises from the application of RE to λC, which gives us the instances of the following schema:

$$(\lambda \bar{p}.\psi)\bar{\varphi} = \psi[\bar{\varphi}/\bar{p}]$$

Some instances of this schema are quite intuitive, such as $(\lambda p.\neg\neg p)q = \neg\neg q$. Others will seem more contentious, such as $(\lambda p.\top)q = \top$. Although the general schema is a relatively strong principle, there are many things to be said in favor of it, as discussed by Dorr (2016, §5) and Goodman (forthcoming).

In general, how plausible is CLASSICISM? We have already seen that many natural and common arguments against similarly coarse-grained views are unpersuasive in the present context. In particular, one cannot use arguments based on attitude ascriptions, as discussed above. For exactly the same reasons, one also cannot appeal to notions like analyticity or apriority in order to argue for a finer individuation of propositions. And as discussed in the supplementary section 3.3, the fine-grained individuation of propositions suggested by the recent notion of metaphysical grounding leads to inconsistency, and so is also not probative.

In contrast to this, modal notions—which were used to argue against EXTENSIONALISM—do not obviously lead to any arguments against CLASSICISM. This is because CLASSICISM is valid on standard possible worlds models, which show strong modal principles to be consistent with CLASSICISM. The same applies to modalities in a broader sense, which include temporal contexts. Other cases are difficult to evaluate; for example, it is far from clear what the correct principles governing deontic terms are, and so difficult to evaluate whether they are consistent with CLASSICISM. They may therefore be set aside for the time being.

CLASSICISM therefore seems viable, and distinguished in terms of simplicity and strength. The following chapters explore it further, and use two extensions of $\mathcal{L}^=$ to motivate two strengthenings of CLASSICISM.

4.6 Supplement: Attitude Ascriptions and Attitudes

This section returns to those who reject the appeal to modal intuitions in arguing against EXTENSIONALISM. We consider two reasons for this. One reason is the idea that modal operators (alongside similar operators, like temporal operators) are really quantifiers, binding hidden variables. In its simplest form, the idea behind this is that an English sentence like "Grass is green" tacitly involves a free variable w, which context determines to be interpreted as standing for the actual world. In "Necessarily, grass is green," "necessarily" serves as a universal quantifier binding this world variable. Such a view is considered by Lewis (1980, §8) and defended by Schaffer (2021). Given this view of English semantics, one might argue that the modal argument against EXTENSIONALISM fails, since the modal data can be accommodated even if every sentential expression is only assigned a truth-value relative to an interpretation of the free variables, including the hidden world variable w.

However, this objection fails to take into account the metasemantic stipulations made in determining the intended interpretation of $\mathcal{L}$. Recall that sentential expressions such as p are stipulated to play a similar role to English sentences like "Grass is green," and sentential operators such as $\Box$ are stipulated to play a similar role to English sentential adverbs like "necessarily," on a suitable interpretation. So, plausibly, if the theory of the syntax and semantics of English sketched in the previous paragraph is correct, then $\mathcal{L}$ will function quite differently from English. After all, since the syntax of $\mathcal{L}$ is defined rigorously at the outset, it cannot later be found to contain hidden variables in sentential expressions like p. Instead, the most natural conjecture is that if the view of English under consideration is correct, then a sentential expression like p which is stipulated to correspond to "Grass is green" will express not a truth-value, but a collection of possible worlds, consisting of the worlds w for which "Grass is green" is true (where w is again the hidden world variable in "Grass is green"). Correspondingly, if $\Box$ is stipulated to correspond to "necessarily," then it will express a property of collections of possible worlds, in particular one which applies to the collection of all possible worlds. The resulting view will therefore not validate EXTENSIONALISM.

It is also worth noting that the simple argument against EXTENSIONALISM could as well be carried out using various other expressions instead of "necessarily," such as temporal expressions like "it is always going to be the case that." In tense logic, such an expression is standardly formalized using a sentential connective G. Analogous to the case of "necessity" just discussed, one might argue that temporal expressions in English are in some sense variable binders; indeed, this is common in formal semantics, following Partee (1973). Even in this case, it is plausible that G can be assigned a non-trivial interpretation formalizing "it is always going to be the case that," and that therefore an argument against EXTENSIONALISM can be formulated using G so interpreted. An interesting question, which will be left open here, is exactly how far these types of arguments can be extended. Potentially, limning the reach of these arguments requires making further metasemantic stipulations, e.g., about which English expressions can be formalized using sentential connectives. For example, it is natural to wonder whether a similar argument can be carried out using expressions concerning location, such as "everywhere" or "on the moon." What is clear is that quantificational phrases such as "everyone" cannot be used in such arguments (cf. Prior 1968), since established practices of formalization use quantifiers in formal logic to regiment such talk instead of sentential operators. These questions are important not only for determining what kinds of arguments against EXTENSIONALISM are available, but also for determining which forms of intensionality are operative in individuating propositions.

A second objection is maybe less plausible, but also harder to refute. This objection rejects the idea that there is any such thing as a circumstantial modality. Instead, the relevant view insists that all readings of modal terms are in some sense epistemic, or otherwise attitudinal, and that there is therefore no way of interpreting $\Box$ on which our intuitions concerning necessity and possibility are probative. In response to this objection, I will argue that although intuitive judgements about attitude ascriptions cannot directly be used to draw conclusions about the individuation of propositions, reflection on attitudes can still give us reasons for rejecting EXTENSIONALISM. The proposal could be regimented more formally in a higher-order language which includes first-order quantifiers, like $\mathcal{L}^*$, but the discussion will be sufficiently straightforward that this won't be necessary.

The fundamental idea behind the response is to distinguish between attitude ascriptions and attitudes. We have seen that we cannot rely in any straightforward way on attitude ascriptions. But there may be attitudes which we can utilize. It will be helpful to consider a particular example. I will use belief, although I could just as well have used one of a number of other notions, such as knowledge. Returning once again to the initial example of PL and OT in the previous chapter, we have seen that we cannot use intuitive judgements about attributions of belief to conclude that there is a relation of belief between agents and propositions which relates a certain agent—who is somewhat of a logical novice—to OT but not to PL. Yet, we would still want to say that this logical novice is in a different doxastic state than a logical expert who we would intuitively describe as believing not just OT but also PL. There is no reason to think that this intuitive difference in doxastic state is only a matter of how we represent reality: there is no reason to doubt that there is a property of being in the same doxastic state as the logical novice, which is not had by the logical expert.

We can go one step further: it might well be that the difference in doxastic states between different agents can be explained in terms of a doxastic relation D between agents and propositions. That is, there might well be such a relation D satisfying the following principle:

(Doxastic States)   Two agents are in the same doxastic state just in case they are related to the same propositions by D.

Furthermore, it may be that any difference in intuitive ascriptions of belief can be explained in terms of actual differences in terms of D. Let me illustrate this idea using PL and OT again. If the sentences PL and OT express the same proposition, then the difference in ascribing belief in PL to the logical expert but not to the novice cannot simply be explained in terms of the former but not the latter standing in the doxastic relation D to the proposition expressed by PL. But it might be possible to explain this difference in terms of the former but not the latter standing in the doxastic relation D to a different proposition, such as the metalinguistic proposition that the sentence PL expresses a truth. I don't mean to propose any kind of metalinguistic semantics of attitude ascriptions; developing such a theory would be far too ambitious an undertaking in the context of this discussion. I merely mean to illustrate how differences in intuitive ascriptions of belief and differences in doxastic states could in principle be explained in terms of differences in the propositions an agent is related to by the underlying doxastic relation D. For more developed accounts of attitude ascriptions which could be used in such explanations, see, e.g., Stalnaker (1999) and Yli-Vakkuri and Hawthorne (2022). On the picture that emerges from the present response, doxastic states track the propositions an agent is related to by D, as do intuitive ascriptions of belief. While doxastic states do so straightforwardly by the principle Doxastic States, intuitive ascriptions of belief track D in ways which are harder to understand.

One might worry that Doxastic States runs into difficulties similar to those we have seen to arise from intuitive belief ascriptions. But as long as there is a sufficient number of propositions, there will be some relation D satisfying DOXASTIC STATES: roughly, as long as the number of inhabited doxastic states is no greater than 2^k, where k is the number of propositions, there will be a map from inhabited doxastic states to collections of propositions which is injective (i.e., which maps distinct states to distinct collections). If we let D relate an agent to a proposition just in case the proposition is in the collection to which the doxastic state of the agent is mapped, DOXASTIC STATES is guaranteed to be satisfied. Conversely, for DOXASTIC STATES to be satisfied, the number of inhabited doxastic states may not be greater than 2^k, where k is the number of propositions. With the plausible assumption that there is no finite number which limits how many inhabited doxastic states there could be, this means that there is no finite number which limits how many propositions there could be. Section 5.5 will argue that it is a necessary matter what propositions there are. It follows that there are infinitely many propositions. Thus, the appeal to DOXASTIC STATES provides us with an alternative argument for the claim ∞, which schematically expresses that there are infinitely many propositions, and therefore against EXTENSIONALISM.

A fuller development of this argument would require fleshing out the notion of a doxastic state, and arguing that there are infinitely many doxastic states. The idea is that this can be done without relying on attitude ascription, in the context of a more general theory of cognitive agency; see Stalnaker (1984) for an account which could be appealed to at this point. The argument is therefore naturally seen as somewhat schematic, dependent on the underlying theory of cognitive agency. Plausibly, such a theory will come with the notion of a doxastic state. In fact, this theory might come directly with a relation D satisfying DOXASTIC STATES, in which case the argument against EXTENSIONALISM could be simplified, and stated directly in terms of D. Here, we only assume the weaker requirement of providing a notion of a doxastic state, which conforms to the intuitive judgement that there are infinitely many doxastic states.

Assume we are given such a theory of cognitive agency, which provides us with an independent grip on the notion of a doxastic state. Why should we think that DOXASTIC STATES is true? One reason is that the metasemantics of propositional quantification involves the constraint of conforming to talk of propositions in English—albeit imperfectly, to allow for the truth of the logical principles which have been stipulated to be true. Especially in a philosophical context, talk of propositions often occurs in discussions of propositional attitudes, and so there is good reason to think that the metasemantic stipulations already laid down will push the interpretation of propositional quantifiers to involve propositional attitudes in some non-trivial way, although guided by the requirement to conform to the stipulative logical principles. If naive judgements about attitude ascriptions conflict with these logical requirements, but the weaker connections imposed by DOXASTIC STATES do not conflict with them, then it is not implausible that the latter but not the former end up being validated. Alternatively, we might ensure the truth of DOXASTIC STATES simply by adding the requirement of satisfying it as a further metasemantic principle.

4.7 Supplement: Identity

This supplementary section shows that there are two formulas involving identity as the only connective which are individually, but not jointly, consistent in $\vdash^=$. This claim is witnessed by the following pair:

(I1) $(p = p) = (q = q)$

(I2) $(p = (p = p)) = (q = (q = q))$

In fact, these are not just individually consistent, but consistent with the equational theory of section 4.3.

Proposition 4.9. I1 *and* I2 *are individually consistent in* $\vdash^=$ *with the set of* $\mathcal{L}^{BQ}$-*equations* $\varphi = \psi$ *such that* $\vdash^= \varphi \leftrightarrow \psi$.

Proof. For I1, consider any standard possible worlds model $\mathfrak{M}$. First, by an induction on the length of derivations, it follows from Lemmas 2.10 and 2.12 that every theorem of $\vdash^=$ is valid on $\mathfrak{M}$. Second, if $\vdash^= \varphi \leftrightarrow \psi$, then it follows by simplicity that $\varphi = \psi$ is valid on $\mathfrak{M}$. Third, simplicity also guarantees that $p = p$ and $q = q$ are interpreted as the set of all worlds W on any assignment function, and so that I1 is valid on $\mathfrak{M}$. The consistency claim follows from these three observations by Lemma 2.8.

For I2, let $\mathfrak{M} = \langle W, D, V \rangle$ be a full possible worlds model based on a two-element set $W = \{w, v\}$ such that V interprets $=$ as follows, for all $d, e \in P_0$:

$$V(=)(\langle d, e \rangle) = \begin{cases} W & \text{if } d = e \neq W \\ \{w\} & \text{if } d = e = W \\ \varnothing & \text{if } d \neq e \end{cases}$$

First, a routine inspection of cases shows that RI and LL are w-valid. By an induction on the length of derivations, it follows from this with Lemmas 2.10 and 2.12 (2) that every theorem of $\vdash^=$ is w-valid. Second, let $\varphi = \psi$ be an $\mathcal{L}^{BQ}$-equation such that $\vdash^= \varphi \leftrightarrow \psi$. By Proposition 4.7, $\vdash \varphi \leftrightarrow \varphi^B$ and $\vdash \psi \leftrightarrow \psi^B$. So $\vdash^= \varphi^B \leftrightarrow \psi^B$. As noted above, among $\mathcal{L}^B$-formulas, provability in $\vdash^=$ coincides with being a tautology, whence $\vdash \varphi^B \leftrightarrow \psi^B$. By Lemmas 2.10 and 2.12 (2), every theorem of $\vdash$ is valid on $\mathfrak{M}$, so for every assignment function a, $[\![\varphi]\!]_a = [\![\varphi^B]\!]_a = [\![\psi^B]\!]_a = [\![\psi]\!]_a$. Thus $\varphi = \psi$ is w-valid. Third, I2 is valid on $\mathfrak{M}$. An inspection of cases shows that $d \neq V(=)(\langle d, d \rangle)$, for all $d \subseteq W$. It follows from this that $[\![p = (p = p)]\!]_a = \varnothing$, and so that I2 is valid on $\mathfrak{M}$. The consistency claim follows from these three observations by Lemma 2.8. □

A deduction in $\vdash^=$ shows that I1 and I2 are incompatible:

Proposition 4.10. I1 *and* I2 *are jointly inconsistent.*

Proof. By a deduction in $\vdash^=$:

(1)	$(p = p) = (\top = \top)$	I1, UG, UI
(2)	$(q = q) = (\top = \top)$	I1, UG, UI
(3)	$(p = (\top = \top)) = (q = (\top = \top))$	I2, 1, 2, LL
(4)	$((\top = \top) = (\top = \top)) = (\bot = (\top = \top))$	3, UG, UI
(5)	$((\top = \top) = (\top = \top)) \rightarrow (\bot = (\top = \top))$	4, LL
(6)	$(\top = \top) = (\top = \top)$	RI
(7)	$\bot = (\top = \top)$	5, 6
(8)	$(\top = \top) \rightarrow \bot$	7, LL
(9)	$\top = \top$	RI
(10)	$\bot$	8, 9

□

4.8 Supplement: Schema vs Rule

This supplementary section confirms the claim made above, that CLASSICISM is stronger than the EQUIVALENCE SCHEMA in $\vdash^=$, as shown by the following sentence:

$$PN := \forall p(\exists q(q = p) = \top) = \top$$

Anticipating the discussion of necessity in the next chapter, this sentence can be understood as stating that it is necessary what propositions there are, which is called PROPOSITIONAL NECESSITISM in (Fritz 2016).

Since $\exists q(q = p) = \top$ is an instance of ES, one easily establishes that the biconditional corresponding to PN follows from ES. That is:

$$ES \vdash^= \forall p(\exists q(q = p) = \top) \leftrightarrow \top$$

Similarly, it is straightforward to show that PN is contained in CLASSICISM, as noted in the next result:

Proposition 4.11. PN $\in$ CLASSICISM.

Proof. By a deduction in $\vdash^= + $ RE:

(1)	$\exists q(q = p)$	$\vdash^=$
(2)	$\exists q(q = p) = \top$	1, RE
(3)	$\forall p(\exists q(q = p) = \top)$	2, UG
(4)	$\forall p(\exists q(q = p) = \top) = \top$	3, RE $\qquad\square$

It therefore only remains to show that ES $\nvdash^=$ PN, i.e., that $\neg$PN is consistent with ES in $\vdash^=$. This can be shown using a possible worlds model. The basic idea behind the model construction is to have three possible worlds w, v, and u. All connectives are interpreted as expected, except for quantifiers in world w: there, they are restricted to entities which do not distinguish between v and u. In particular, every proposition which exists at w exists at every world, whereas the proposition true only in v exists at v and u but not at w. With this, the falsity of PN in w follows. It then suffices to show that every instance of ES and every theorem of $\vdash^=$ is w-valid.

To start, let W be a three-element set $\{w, v, u\}$. Any permutation f of W can be extended to P_n, by defining $f.b \in P_0$ and $f.o \in P_n$, for any $n > 0$, $b \in P_0$, and $o \in P_n$, as follows:

$$f.b := \{f(w) : w \in b\}$$
$$f.o := \langle b_1, \ldots, b_n \rangle \in P_0^n \mapsto f.o(\langle f^{-1}.b_1, \ldots, f^{-1}.b_n \rangle)$$

Let (vu) be the permutation of W which maps w to itself, and v and u to each other. Each world $x \in W$ can now be assigned a domain D_n^x for any type $n \in \mathbb{N}$:

$$D_n^w := \{o \in P_n : (vu).o = o\}$$
$$D_n^v = D_n^u := P_n$$

Based on W and D, let $\mathfrak{M}$ be a simple possible worlds model $\langle W, D, V \rangle$. The following lemmas show that this model has the required properties:

Lemma 4.12. PN *is not w-valid.*

Proof. Let a be any assignment function. Since $\mathfrak{M}, w, a[\{v\}/p] \not\models \exists q(q = p)$, it follows that $\mathfrak{M}, v, a \not\models \forall p(\exists q(q = p) = \top)$. So PN is not w-valid. $\square$

Lemma 4.13. *If a is a w-assignment, then for any expression ε, $[\![\varepsilon]\!]_a \in D_n^w$.*

Proof. For every assignment function a, let a^* be the function mapping every variable x to $(vu).a(x)$. An induction on the complexity of ε shows that for every assignment function a, $(vu).[\![\varepsilon]\!]_a = [\![\varepsilon]\!]_{a^*}$. The claim follows, as $a = a^*$ for every w-assignment a. $\square$

Lemma 4.14. *If $\vdash^= \varphi$, then $\mathfrak{M}, x, a \models \varphi$ for every $x \in W$ and w-assignment a.*

Proof. By induction on the length of proofs. The case of UI follows with Lemmas 2.9 (2) and 4.13; the remaining cases are established in Lemmas 2.10 and 2.12 (2). $\square$

Lemma 4.15. *Every instance of* ES *is w-valid.*

Proof. Let a be a w-valuation, and assume $\vdash^= \varphi \leftrightarrow \psi$. By Lemma 4.14, $\mathfrak{M}, x, a \models \varphi \leftrightarrow \psi$ for all $x \in W$. So $\varphi = \psi$ is w-valid. □

Proposition 4.16. ES $\nvdash^=$ PN.

Proof. By the preceding lemmas using Lemma 2.8. □

Propositions 4.11 and 4.16 show that CLASSICISM is stronger than ES. It is worth noting that this result is dependent on the particular formal system $\vdash^=$ employed here; in the setting of Bacon and Dorr (forthcoming), which provides a fuller higher-order language along the lines of $\mathcal{L}^*$ rather than the more restrictive language $\mathcal{L}$, schema and rule turn out to be interchangeable.

PART III
NECESSITY

5

Metaphysical Necessity

5.1 Classicism and Intensionalism

CLASSICISM may be an unfamiliar view. But its characteristic feature—
of entailing that two formulas express the same proposition if they are
provably equivalent (in a suitable deductive system)—is familiar from a
more common view. According to this more familiar view, necessarily
equivalent propositions are identical. Call this view INTENSIONALISM.
Provable formulas are normally taken to express necessary propositions.
From this, it follows that provably equivalent formulas express necessarily equivalent propositions. With INTENSIONALISM, it follows that
provably equivalent formulas express the same proposition.

With □ expressing necessity, INTENSIONALISM can be stated as the
following principle of $\mathcal{L}^=$:

(INTENSIONALISM) $\Box(p \leftrightarrow q) \rightarrow (p = q)$

With standard deductive principles of modal logic, many of the distinctive consequences of CLASSICISM can be obtained from INTENSION-
ALISM. For example, if the rule of necessitation (according to which
$\Box\varphi$ can be derived from φ) is added to $\vdash^=$, then every instance of the
EQUIVALENCE SCHEMA can be derived using INTENSIONALISM. Never-
theless, the statement of INTENSIONALISM is importantly different from
the statement of CLASSICISM: whereas the former is specified using a
single axiom, the latter uses the RULE OF EQUIVALENCE to characterize its
members; furthermore, whereas this formulation of CLASSICISM makes
use of a deductive notion of derivability, the statement of INTENSIONAL-
ISM involves a primitive notion of necessity.

The appeal to a notion of necessity in formulating INTENSIONALISM
is not unproblematic, since modal terms in English are highly context-
sensitive. A view like INTENSIONALISM is most commonly understood as

The Foundations of Modality: From Propositions to Possible Worlds. Peter Fritz, Oxford University Press.
© Peter Fritz 2023. DOI: 10.1093/oso/9780192870025.003.0006

pertaining to a special modality of metaphysical necessity. Although the notion of metaphysical necessity is widely appealed to in metaphysics, there is also considerable skepticism concerning the notion. Such skepticism does not require a general skepticism with respect to modality: even if there are various modalities, metaphysical theorizing may not distinguish any particular modality as metaphysical necessity.

This chapter shows how CLASSICISM can be used to make progress on these issues. First, I will argue that if there is a necessity which is broadest (in a sense to be made precise), then this determines the behavior of metaphysical necessity. Second, I will use CLASSICISM to argue that there is indeed a broadest necessity, namely *being* $\top$, the modality expressed by the operator $\lambda p.p = \top$. Finally, it will be shown that INTENSIONALISM is true for this modality. With λC, this amounts to showing that $p = q$ whenever $p \leftrightarrow q$ is the tautologous proposition $\top$. This is in fact easily done, and it will be useful to establish it right away:

Lemma 5.1. CLASSICISM $\vdash^= ((p \leftrightarrow q) = \top) \to (p = q)$

Proof. Assume $(p \leftrightarrow q) = \top$. Relying on equations which follow with CLASSICISM, the following chain of identities can be proven:

$$p = (\top \wedge p) = ((p \leftrightarrow q) \wedge p) = (p \wedge q) = (q \wedge (p \leftrightarrow q)) = (q \wedge \top) = q.$$

$\square$

5.2 Metaphysical Necessity

Talk of metaphysical necessity is ubiquitous in contemporary metaphysics. Often, the qualifier "metaphysical" is omitted, and disputes are simply phrased in the modal terms of English. But in order for disputes in modal metaphysics not to be in danger of being merely verbal, it is important that some special notion of necessity is at play, even if this is not always made explicit. This is because modal terms are ordinarily highly dependent on context. Modal terms may also be ambiguous, but the distinction between context-dependence and ambiguity will not make any difference in the following, so we can set ambiguity aside.

We already noted the context-dependence of modal terms in section 4.1. We saw that these terms have different readings, which include epistemic and deontic ones. Even settling on, e.g., an epistemic reading of modal terms, which is concerned with knowledge, the meaning of a modal term is still dependent on whose knowledge is at issue. We noted that Kripke (1980 [1972]) argued for a metaphysical reading of modal terms. This reading is meant to be concerned with the world itself, rather than anything like someone's epistemic access to it, or permissibility according to some rules. But analogous to the case of epistemic readings, there is no reason to think that this qualification specifies a single interpretation of modal terms, rather than a whole class of them. Above, we called these readings "circumstantial"; Williamson (2016) develops a general theory of such modalities, calling them "objective."

This multiplicity of modalities has led a number of authors to question the assumption that there is a distinguished notion of metaphysical necessity which plays a central role in metaphysics, including Field (1989, p. 39); Sider (2011, ch. 12); Nolan (2011); Priest (2021); and Clarke-Doane (2021). Even one of the most influential philosophers working on modal metaphysics, Fine (2005, p. 7), who widely appeals to metaphysical necessity in his work, acknowledges the concern:

> What is it for a truth to hold or for a feature to be had of *metaphysical* necessity? Philosophers have not given this question the attention it deserves; they have simply taken for granted that there is a single coherent notion that goes by this name.

The worry can be sharpened. One may not just fail to see any argument *for* the existence of a distinguished notion of metaphysical necessity, but also see reasons *against* the existence of such a distinguished notion. For example, modal puzzles of recombination such as the one described in Fritz (2017) may push one to think that degrees of necessity are in some sense indefinitely extensible, a view which is articulated and defended in Rayo (2020); see also Clarke-Doane (2019, pp. 283–284).

However, using Classicism, a strong argument can be made for the existence of a distinguished notion of metaphysical necessity. The starting point of this argument is the observation that the widespread talk of metaphysical necessity in contemporary metaphysics owes much to a famous series of three lectures given at Princeton University by Kripke (1980 [1972]). There, Kripke mainly talks just about "necessity," although he explicitly states being concerned with a notion "of metaphysics, in some (I hope) nonpejorative sense" (pp. 35–36). Furthermore, at the end of the second lecture, he briefly addresses the question which (metaphysical) notion of necessity he is concerned with (p. 99):

> here of course I don't mean just physically necessary, but necessary in the highest degree—whatever that means. (Physical necessity, *might* turn out to be necessity in the highest degree. But that's a question which I don't which to prejudge. At least for this sort of example, it might be that when something's physically necessary, it always is necessary *tout court*.)

Applying Kripke's causal chain theory of metasemantics put forth in the same lectures, it is plausible that what current uses of "metaphysical necessity" express is determined by what Kripke talked about himself. The main criterion he offered for what he was talking about was being necessity in the highest degree (or, as he alternatively puts it, necessity *tout court*). It follows from this that current uses of "metaphysical necessity" also express necessity in the highest degree, assuming there is such a notion. This conclusion coheres with what many writers after Kripke have said about metaphysical necessity, including van Inwagen (1998, p. 72); Stalnaker (2003, p. 203); Rosen (2006); and Williamson (2016).

Given the assumption that there is a notion of necessity in the highest degree, these metasemantic considerations go a long way toward allaying worries about talk of metaphysical necessity. It does remain to justify the assumption that there is a notion of necessity in the highest degree. But it will now be shown that this assumption follows from CLASSICISM.

5.3 The Broadest Necessity

Terminologically, Kripke's talk of *degrees of necessity* is awkward to use in comparing necessities. Following Bacon (2018a), one necessity being of a higher degree than another will therefore from now on be rephrased as the former being *broader* than the latter. The claim under consideration is therefore that there is a broadest necessity. Two questions need to be addressed to assess this claim. First, what is it to be a necessity? Second, what is it for one necessity to be at least as broad as another?

Starting with the question what it is to be a necessity, some unproblematic requirements are easily identified. For example, any proposition expressed by any theorem of classical propositional logic should be necessary, whatever notion of necessity is considered. According to CLASSICISM, every such sentence expresses the same tautological proposition ⊤. The requirement just mentioned can therefore be formulated by stating that $m\top$ for every necessity m.

Although some requirements for being a necessity are easily formulated, it is not so easy to give necessary *and sufficient* conditions. Bacon (2018a) suggests that a property of propositions m is a necessity if and only if every property n which applies to ⊤ applies also to $m\top$. In $\mathcal{L}$, this can be stated as follows:

$$\forall n(n\top \rightarrow nm\top)$$

But this is fairly weak. For example, it does not follow from this (in $\vdash^=$) that every necessity m is factive, in the sense that $\forall p(mp \rightarrow p)$, nor that what is m-necessary is closed under modus ponens, in the sense that $\forall p \forall q(m(p \rightarrow q) \rightarrow (mp \rightarrow mq))$, as is easily demonstrated using possible worlds models. Yet these are natural further requirements on necessities.

Admittedly, although these further requirements are natural, there are also reasons to question them. Consider factivity, by way of example. Thinking of paradigmatic cases like physical necessity or the controversial notion of metaphysical necessity, it may initially seem plausible that if p is necessary, according to any sense of necessity, then p should be the case. But it is less clear that factivity is required when we consider some other cases, such as deontic modalities, as it may be obligatory to bring about p without p being the case. This raises the question whether deontic modalities should count as necessities, in the relevant sense.

One might attempt to make progress on these difficulties by restricting oneself to objective modalities, in Williamson's terms. It could then be argued that deontic modalities are not objective. But this only leads to the further question what it is for a necessity to be objective. The linguistic literature, at least, inspires little confidence that the different readings of English modal terms divide neatly into a small finite number of categories, as there appear to be as many taxonomies as authors (see again Portner (2009, §4.1)). And Williamson's own distinction between objective and non-objective modalities in fact fits uneasily in the present discussion: Williamson (2016, p. 454) considers a crucial difference between objective and epistemic modalities to be the fact that the latter are, but the former are not, sensitive to the guises in which relevant entities are presented. But in the present discussion, a modality is a property of propositions, and these cannot be sensitive to guises, since LL is true by stipulation. The problematic guise-sensitive epistemic readings of English modal terms have therefore already been eliminated through the use of $\mathcal{L}$. Williamson (2016) says little about how deontic and other kinds of modalities may be distinguished from objective modalities. In the present context, it is therefore doubtful that much light will be shed on the relevant notion of a necessity by introducing a notion of objectivity.

Trying to give a complete account of what it takes to be an (objective) necessity therefore quickly leads to some very difficult questions. And on reflection, it is doubtful that this should be regarded as a substantial question. Plausibly, notions such as that of being a necessity are theoretical notions of philosophy, and without further stipulations, many disputes involving them, such as whether all necessities are factive, are likely to be verbal.

Fortunately, it turns out that for present purposes, it is not necessary to settle on a unique notion of what it takes to be a necessity. Rather, it suffices to impose two general constraints, which the most natural ways of understanding the notion of a necessity in philosophy will all satisfy. Both of them involve the modality $\square$ of being the tautologous proposition, defined as follows:

$$\square := \lambda p.p = \top$$

This modality will play an important role. In fact, I will argue that $\square$ is the broadest, and thus metaphysical, necessity. To be clear, $\square$ is here used as a metalinguistic abbreviation, in order to specify expressions of $\mathcal{L}^=$ involving $\lambda p.p = \top$. Later on, we return to using $\square$ as a unary operator constant, although then it will be interpreted as expressing the modality $\lambda p.p = \top$.

The first constraint on necessities requires $\Box$ to be a necessity. $\Box$ applies to $\top$, since $\top$ is $\top$. $\Box$ is also factive, since $\top$ is the case. More generally, it will be shown below that $\Box$ satisfies a very strong and attractive modal logic. Being the tautologous proposition also fits well with one way in which various authors have explicated the notion of necessity they are concerned with. For example, Lewis and Langford (1959 [1932], pp. 248–249) state that p is necessary just in case "[t]he denial of p is not self-consistent." Consistency is a property which sentences have relative to deductive systems, so this account appears to conflate the distinction between sentences and the propositions they express. But given CLASSICISM, every contradiction (every formula whose negation is provable in $\vdash^=$) expresses the same proposition $\bot$. The denial of p not being self-consistent can thus naturally be understood as $\neg p$ being $\bot$, which, given CLASSICISM, is equivalent to p being $\top$. More explicit definitions of $\Box$ in terms of $\top$ along the present lines can be found already in Church (1951); Cresswell (1965); and Suszko (1971).

The second constraint on necessities is a generalization of the constraint mentioned earlier, of applying to the tautologous proposition. As it stands, this is a very weak constraint. Consider the property m of being asserted by a particular speaker. Assuming the speaker does in fact assert the tautology, e.g., by uttering some theorem of classical propositional logic, m satisfies the condition of applying to $\top$. But it might well be *possible* for the speaker never to assert the tautology, in some relevant sense of "possible." That is, there might well be a notion of necessity n such that n does not apply to $m\top$. And this is a good reason not to recognize m as a necessity, on a natural way of thinking of necessities. This line of argument motivates the constraint that if m and n are necessities, then it should be the case that $mn\top$. By the first constraint, $\Box$ is a necessity. So in particular, it should be $\Box$-necessary that $\top$ is m-necessary. That is, if m is a necessity, then $\Box m\top$. This is the second constraint.

To state these constraints formally, a way of ascribing to m being a necessity is required. So, let Nec(m) be a formula which does so. In a higher-order expansion of $\mathcal{L}^=$, Nec could simply be taken to be a constant taking unary operators as arguments. Nec(m) could also be a formula of $\mathcal{L}^=$. For present purposes, these choices make no difference, so this aspect will intentionally be left underspecified. Combining the two constraints mentioned above leads to the following rudimentary theory of necessities:

Definition 5.2. *Let* NT *be the theory consisting of the following axioms:*

(N1) Nec($\Box$)

(N2) Nec(m) $\rightarrow$ $\Box m\top$

Before moving on to regimenting the notion of broadness among necessities, it is useful to note that given the theory NT just proposed, $\Box p$ implies mp, for any necessity m and proposition p. For assume $\Box p$; then $p = \top$. And by N2, $\Box m\top$, whence $m\top$. Thus mp. So, if something is necessary according to $\Box$, then it is necessary according to every notion of necessity. And conversely, since by N1, $\Box$ is a necessity, if something is necessary according to every notion of necessity, then it is $\Box$-necessary as well. So the propositions which are $\Box$-necessary are just those which are necessary according to every notion of necessity.

Having put constraints on what it takes to be a necessity, consider now the notion of one necessity m being at least as broad as another necessity n. Clearly, this requires that any m-necessary proposition p is also n-necessary. That is:

$$\forall p(mp \rightarrow np)$$

But this is again a somewhat weak constraint. The reason is that it does not rule out the existence of a proposition p which is *possibly* m-necessary without being n-necessary. So, for any given proposition p, it should not merely be the case that $mp \rightarrow np$ is true, but this conditional should be necessary, on every notion of necessity. Conversely, if for every proposition p, it is necessary, on every notion of necessity, that $mp \rightarrow np$, then m is plausibly at least as broad as n. As just seen, to be necessary on every notion of necessity is equivalent to being $\Box$-necessary. So a necessity m being at least as broad as a necessity n can plausibly be defined as $mp \rightarrow np$ being $\Box$-necessary, for every proposition p. So, μ being *at least as broad as* ν, written $\mu \sqsubseteq \nu$, can be defined as follows:

$$\mu \sqsubseteq \nu := \forall p \Box (\mu p \to \nu p)$$

One might wonder if this condition could not plausibly be strengthened by adding additional $\Box$ operators before or after the universal quantifier. However, it follows from results below that given Classicism, such variant conditions are equivalent to the one given here.

It can now be proven, as claimed earlier, that $\Box$ is broadest among the necessities. That is, it follows from the rudimentary theory of necessities NT and Classicism, first, that $\Box$ is a necessity and, second, that $\Box$ is at least as broad as any necessity m. The first claim is simply N1. The second is established as follows:

Proposition 5.3. Classicism, N2 $\vdash^= \mathrm{Nec}(m) \to \Box \sqsubseteq m$.

Proof. The following indicates a deduction witnessing the claim. Lines 2 and 3 are obtained using standard modal inferences, which will be shown below to be available for $\Box$ given Classicism.

(1) $m\top \to (p = \top \to mp)$ \hspace{2em} LL
(2) $\Box m\top \to \Box(\Box p \to mp)$ \hspace{1.5em} 1
(3) $\Box m\top \to (\Box \sqsubseteq m)$ \hspace{2.5em} 2
(4) N2 $\to (\mathrm{Nec}(m) \to \Box \sqsubseteq m)$ \hspace{0.5em} 3

$\Box$

So far, it has been argued—on the assumption of Classicism—that $\Box$ is a necessity which is at least as broad as any necessity. In this sense, $\Box$ is *a* broadest necessity. This raises the question whether $\Box$ is *the* broadest necessity. If there is another broadest necessity $\Box'$, then:

$$\Box \sqsubseteq \Box' \wedge \Box' \sqsubseteq \Box$$

Given the definition of $\sqsubseteq$, it follows that for any proposition p, $(\Box p \to \Box' p) = \top$ and $(\Box' p \to \Box p) = \top$, whence with Classicism, $(\Box p \leftrightarrow \Box' p) = \top$. By Lemma 5.1, it follows that $\Box p = \Box' p$. Thus $\forall p(\Box p = \Box' p)$, and so:

$$\Box \equiv \Box'$$

As far as modalities can be compared in $\mathcal{L}^=$, $\Box$ is therefore the unique broadest necessity.

If $\Box$ is the unique broadest necessity, then Kripke's characterization of metaphysical necessity as the broadest necessity uniquely picks out $\Box$ as metaphysical necessity. But even if $\Box$ is merely *a* broadest necessity, Kripke's characterization still entails that a proposition is metaphysically necessary just in case it is $\Box$, and indeed that for a proposition to be metaphysically necessary is for it to be $\Box$. Even in the latter case, the skeptical arguments against metaphysical necessity—which center on the problem of delineating what counts as metaphysically necessary— are answered. For simplicity, the following assumes in informal discussions that modalities are individuated according to their applicative behavior, and so that $\Box$ *is* metaphysical necessity.

5.4 An Objection

Before further investigating the theory of metaphysical necessity we have arrived at, it is worth addressing an objection. The objection takes issue with the claim that metaphysical necessity is the broadest necessity. According to this objection, there are necessities *n* and propositions *p* such that *p* is metaphysically necessary without being *n*-necessary. A common instance of this objection notes that although it is metaphysically necessary that Hesperus is Phosphorus, it is neither *a priori* nor logically necessary that Hesperus is Phosphorus; see, e.g., Hale (1996, pp. 98–99); Rumfitt (2010, p. 46); Clarke-Doane (2021, §2); and Mallozzi (forthcoming). However, as Bacon (2018a, pp. 736–737) notes, such cases are problematic, due to the fact that "*a priori*" and "logically necessary" must be conceived either as expressing properties of sentences rather than propositions, or as creating opaque contexts. In the present setting, this means that such contexts must be discounted, just as attitude ascriptions were discounted in Chapter 3. A more detailed discussion of this point can be found in Dorr et al. (2021).

Bacon (2018a, pp. 743–745) nevertheless argues that there are examples of necessities which provide counterexamples to the claim that metaphysical necessity is the broadest necessity. Bacon's main examples are the modality of determinacy, as employed in discussions of vagueness, and temporal modalities such as being always true. But both cases are problematic: *Pace* Bacon (2018a,b), it is plausible and widely endorsed that vagueness is a matter of the relation between words and what they express, or at least a matter of representation more generally; see Sorensen (2022, §9). Consequently, determinacy—like apriority and logical necessity—is either (implicitly or explicitly) metalinguistic or it gives rise to opacity. The case from temporal operators rests on the claim that some propositions can be necessary without always being true. But as Dorr and Goodman (2020) argue, this is a questionable claim. It is especially questionable in the present context, where the relevant notion of necessity is not tied to ordinary uses of English modal terms.

The question whether $\Box$ (*being* $\top$, the broadest necessity) is metaphysical necessity is of interest for making sense of existing debates in modal metaphysics. But, looking forward, it may well be most fruitful to set aside talk of metaphysical necessity. Assuming the coarse-grained individuation of propositions of Classicism, $\Box$ is of obvious interest as the broadest necessity, capturing what others, including Hale (1996), have termed *absolute* necessity. Presumably, there is a wide range of necessities which are narrower than $\Box$. The question whether there are any particular cases of such narrower necessities which may play important roles in metaphysics, and philosophy more generally, is obviously important. It is less clear that it is ultimately so important which, if any, of them is metaphysical necessity. In the following, we set these terminological issues aside, and assume that $\Box$ is metaphysical necessity.

5.5 The Logic of Necessity

Having given an account of metaphysical necessity as *being* $\top$, it is natural to ask which principles $\Box$ satisfies according to Classicism. As noted above, Intensionalism follows immediately from Lemma 5.1. The converse of Intensionalism is immediate using LL, so Classicism entails:

$$(p = q) \leftrightarrow \Box(p \leftrightarrow q)$$

And with RE, the corresponding identification holds, so according to CLASSICISM, for propositions to be identical *is* for them to be necessarily equivalent:

$$(p = q) = \Box(p \leftrightarrow q)$$

Another important principle of necessity and identity which can be derived from CLASSICISM is the necessity of identity:

$$p = q \rightarrow \Box(p = q)$$

This follows from $(p = p) = \top$, which is an instance of CLASSICISM, and LL.

Considering such particular principles provides only a very unsystematic picture of the modal principles entailed by CLASSICISM. A somewhat more systematic picture can be provided by considering *purely modal* principles; that is, principles of $\mathcal{L}^=$ in which $=$ occurs only in occurrences of $\Box$. We can provide an informative characterization of such principles using a proof system involving only purely modal principles. This can be made precise by considering $\Box$ alternatively as a unary operator constant, with an intended interpretation as metaphysical necessity. Let $\mathcal{L}^\Box$ be the language $\mathcal{L}$ with such a constant. The relevant principles of $\mathcal{L}^\Box$ can then be specified using the following extension of $\vdash$:

Definition 5.4. *Let* $\vdash^\Box$ *be the proof system extending* $\vdash$ *by the following axioms and rules:*

(K)	$\Box(p \rightarrow q) \rightarrow (\Box p \rightarrow \Box q)$	(N)	$\varphi/\Box\varphi$
(T)	$\Box p \rightarrow p$	(4)	$\Box p \rightarrow \Box\Box p$
(INT-LL)	$\Box(p \leftrightarrow q) \rightarrow (\varphi(p) \rightarrow \varphi(q))$		

Every formula of $\mathcal{L}^\Box$ can naturally be translated into a formula of $\mathcal{L}^=$, by replacing $\Box$, considered as a constant, by $\lambda p.p = \top$ (which above was abbreviated using $\Box$). It can be shown that a formula of $\mathcal{L}^\Box$ is translated into a formula provable in $\vdash^=$ from CLASSICISM if and only if it is provable in $\vdash^\Box$. In this sense, the modal logic of metaphysical necessity, according to CLASSICISM, is precisely characterized by $\vdash^\Box$. Additionally, it can be shown that the correspondence between $\mathcal{L}^\Box$ and $\mathcal{L}^=$ can be inverted, in the following way. Recall that on the present picture, $p = q$ is $\Box(p \leftrightarrow q)$, for any p and q. One can therefore also define a translation from $\mathcal{L}^=$ into $\mathcal{L}^\Box$, which replaces $=$ by $\lambda pq.\Box(p \leftrightarrow q)$. The question is then: which formulas of $\mathcal{L}^=$ are translated into theorems of $\vdash^\Box$? The answer is that those are just the ones provable in $\vdash^=$ from Classicism. Furthermore, the two translations just sketched are mutual inverses, modulo provability in the relevant proof system. A supplementary section 5.6 establishes these claims. In this sense, metaphysical necessity and propositional identity can be seen as interdefinable, and Classicism and $\vdash^\Box$ can be seen as variant presentations of the same theory. These results are closely related to results which were established, in more restricted languages, by Cresswell (1965, 1967) and Suszko (1971), and, in a fuller higher-order logic, by Bacon (2018a).

The results established here entail a number of consequences for the logic of necessity according to Classicism. For example, since $\vdash^\Box$ includes the axioms and rules of a standard proof system of the modal logic S4, it follows that every theorem of S4 holds for metaphysical necessity. This extends to the standard extension of S4 by the principles of elementary quantification theory for propositional quantifiers; a more detailed definition of such a propositionally quantified modal logic can be found in Fritz (forthcoming). By a standard argument of quantified modal logic, which can be found in textbooks like Hughes and Cresswell (1996), this means that the converse of the Barcan formula of Barcan (1946) for propositional quantifiers is derivable. This is the following schematic principle:

(CBF) $\Box \forall p \varphi \rightarrow \forall p \Box \varphi$

Consequently, the same holds for the claim that (necessarily) every proposition is a necessary existent, as stated by the following principle (a version of which was already mentioned in section 4.8):

$$\Box \forall p \Box \exists q (p = q)$$

As discussed by Fritz (2016), this is a non-trivial commitment, a point to which section 10.1 returns.

The theory of metaphysical necessity developed here, in conjunction with the theory CLASSICISM of propositional identity motivated above, therefore leads to a strong logic of necessity, which proves many standard modal principles. There are, however, some standard modal principles which have not yet been mentioned, and which are not obviously entailed. Two such principles concern the necessity of distinctness and the necessity of possibility. These will be considered in the next chapter. The remainder of this chapter consists of two supplementary sections. The first establishes the formal claims made in this section. The second considers whether there is a way of pinning down the interpretation of the counterfactual conditional in metaphysics, analogous to how the interpretation of modal terms in metaphysics has been pinned down in this chapter, and concludes that the prospects for doing so are dim.

5.6 Supplement: Equivalence

This supplementary section proves the equivalence of $\vdash^{\square}$ with CLASSICISM in $\vdash^{=}$ claimed in the previous section. First, the translations between the two languages are defined as follows:

Definition 5.5. *Let* $\cdot^{\dagger} : \mathcal{L}^{=} \to \mathcal{L}^{\square}$ *and* $\cdot^{\ddagger} : \mathcal{L}^{\square} \to \mathcal{L}^{=}$ *be recursive mappings, with the only non-trivial clauses:*

$$=^{\dagger} := \lambda pq.\square(p \leftrightarrow q)$$
$$\square^{\ddagger} := \lambda p.p = \top$$

The results characterizing the correspondence between propositional identity and metaphysical necessity are as follows:

Proposition 5.6. *The following hold for all* $\varphi \in \mathcal{L}^{=}$ *and* $\psi \in \mathcal{L}^{\square}$:

(1) CLASSICISM $\vdash^{=} \varphi \leftrightarrow \varphi^{\dagger\ddagger}$
(2) $\vdash^{\square} \psi \leftrightarrow \psi^{\ddagger\dagger}$
(3) CLASSICISM $\vdash^{=} \varphi$ *iff* $\vdash^{\square} \varphi^{\dagger}$
(4) $\vdash^{\square} \psi$ *iff* CLASSICISM $\vdash^{=} \psi^{\ddagger}$

Proof of (1). By induction on the complexity of formulas. Since CLASSICISM is closed under RE, the cases other than = follow by IH (induction hypothesis). For =, note that $(\varphi = \psi)^{\dagger\ddagger}$ is equivalent, via λC, to $(\varphi^{\dagger\ddagger} \leftrightarrow \psi^{\dagger\ddagger}) = \top$. By IH, this is equivalent to $(\varphi \leftrightarrow \psi) = \top$. Thus the claim follows by Lemma 5.1. $\qquad\qquad\square$

Proof of (2). By induction on the complexity of formulas. Using INT-LL, the cases other than $\square$ follow by IH. For $\square$, note that $(\square\varphi)^{\ddagger\dagger}$ is equivalent, via λC, to $\square(\varphi^{\ddagger\dagger} \leftrightarrow \top)$. By IH, N and INT-LL, this is equivalent to $\square(\varphi \leftrightarrow \top)$. A routine argument in S4 shows that this is equivalent to $\square\varphi$. $\qquad\qquad\square$

To show (3) and (4), it suffices to establish the left to right directions of these two claims; the right to left directions follow from these with (1) and (2).

Proof of (3$\Rightarrow$). By induction on the length of proofs in $\vdash^= +$RE. The cases of the principles of $\vdash$ are immediate, using the fact that these principles are also included in $\vdash^\square$.

RI: $(p = p)^\dagger$ is equivalent, via λC, to $\square(p \leftrightarrow p)$, which is straightforward to derive in $\vdash^\square$.

LL: Any instance of LL is mapped by $\cdot^\dagger$ to a formula which is equivalent, via λC, to an instance of INT-LL.

RE: Assume $\varphi \leftrightarrow \psi$ is provable in $\vdash^= +$ RE, whence by RE, $\varphi = \psi$ is as well. By IH, it follows that $\varphi^\dagger \leftrightarrow \psi^\dagger$ is derivable in $\vdash^\square$. An application of N yields $\square(\varphi^\dagger \leftrightarrow \psi^\dagger)$, which is equivalent, via λC, to $(\varphi = \psi)^\dagger$, as required. $\qquad\qquad\square$

Proof of (4$\Rightarrow$). By induction on the length of proofs in $\vdash^\square$. As in 3$\Rightarrow$, the cases of principles of $\vdash$ are immediate, using the fact that these principles are also included in $\vdash^=$.

K: $K^\ddagger$ is equivalent, via λC, to the following formula:

$$((p \rightarrow q) = \top) \rightarrow ((p = \top) \rightarrow (q = \top))$$

This can be derived in $\vdash^= + \mathrm{RE}$, as the following sketch of a derivation shows. Assume $(p \to q) = \top$ and $p = \top$. With RE, the following chain of equations follows:

$$q = (q \wedge \top) = (q \wedge p) = (p \wedge (p \to q)) = (\top \wedge \top) = \top$$

N: Assume φ is provable in $\vdash^{\square}$, whence by N, $\square\varphi$ is as well. By IH, it follows that $\varphi^{\ddagger}$, and so $\varphi^{\ddagger} \leftrightarrow \top$, are derivable in $\vdash^= + \mathrm{RE}$. An application of RE yields $\varphi^{\ddagger} = \top$, whence $(\square\varphi)^{\ddagger}$ using $\lambda\mathrm{C}$, as required.

T: $\mathrm{T}^{\ddagger}$ is equivalent, via $\lambda\mathrm{C}$, to $(p = \top) \to p$, which is straightforward to derive in $\vdash^=$ using LL.

4: $4^{\ddagger}$ is equivalent, via $\lambda\mathrm{C}$, to $(p = \top) \to ((p = \top) = \top)$. By LL, this follows from $(\top = \top) = \top$, which itself follows from RI using RE.

INT-LL: Any instance of INT-LL is mapped by $\cdot^{\dagger}$ to a formula which follows, using $\lambda\mathrm{C}$ and Lemma 5.1, from a corresponding instance of LL. □

5.7 Supplement: Counterfactuals

In addition to theorizing in terms of necessity and possibility, work in metaphysics over the last fifty years contains many appeals to counterfactual conditionals, such as claims of the form "had φ been the case, ψ would have been the case." Such conditionals were influentially discussed by Stalnaker (1968) and Lewis (1973).

As in the case of modal terms like "necessary" and "possible," it is widely acknowledged that counterfactual conditionals in English are highly context-sensitive. And as in the case of these modal terms, many uses of counterfactuals in substantive metaphysical theorizing therefore rely on a way of resolving the context-dependence of counterfactuals. Given the way in which CLASSICISM allows for a satisfying resolution of this dependence and provides a definition of metaphysical necessity in terms of identity, it is natural to wonder whether the use of counterfactuals in metaphysics can similarly be put on a firmer footing.

Since metaphysical necessity was singled out as the broadest necessity, one might try to single out a distinguished "metaphysical" counterfactual conditional as the broadest counterfactual conditional. Analogous to before, this requires an account of the relevant notion of broadness, and an account of the relevant notion of being a counterfactual conditional. It is relatively straightforward to adapt the standard of broadness. Recall that m was said to be at least as broad as n just in case every proposition is, ($\Box$-)necessarily, m-necessary only if it is n-necessary. Analogously, it is natural to think that one counterfactual conditional is at least as broad as another just in case for any arguments, necessarily, an instance of the former holds only if the corresponding instance of the latter holds. That is, a relation of comparative broadness among counterfactual conditionals can be defined as follows:

$$\Box\!\!\rightarrow\ \sqsubseteq > \ := \forall p \forall q \Box((p \Box\!\!\rightarrow q) \rightarrow (p > q))$$

As in the case of necessities, it is hard to say which binary modalities should count as counterfactual conditionals. But again, it turns out that much can be concluded merely from some constraints on what may count as a counterfactual conditional. So, assume that for every binary sentential operator $\Box\!\!\rightarrow$, there is a sentence $CC(\Box\!\!\rightarrow)$ stating that $\Box\!\!\rightarrow$ is a counterfactual conditional. Again, this might be a formula of $\mathcal{L}^=$, or CC might be treated as a constant in a higher-order extension of $\mathcal{L}^=$.

First, among the constraints on CC, it is natural to include standard principles of the logic of counterfactuals. An example of this is the principle that a metaphysically strict implication metaphysically strictly implies the corresponding counterfactual. For example, this principle is easily seen to be valid on the model theory for counterfactuals of Lewis (1973). This leads to the following constraint:

(C1) $CC(\Box\!\!\rightarrow) \rightarrow \Box(\Box(p \rightarrow q) \rightarrow (p \Box\!\!\rightarrow q))$

Second, according to a popular theory of counterfactuals, counterfactuals can be understood as strict conditionals, with the relevant modality being determined by context; see Starr (2019, §2.2). On this view, apparent counterexamples to the view that counterfactuals are strict conditionals are explained away by context-sensitivity. The context-dependence of counterfactuals therefore goes hand-in-hand with the context-dependence of modal terms. Since metaphysical necessity is one available reading of modal terms, it follows from this that the metaphysically strict conditional is an example of a counterfactual conditional:

(C2) $CC(\lambda pq.\Box(p \to q))$

This is plausible even without assuming that counterfactuals can in general be understood as strict conditionals, since it is compatible with there being readings of counterfactuals which fall outside the scope of this theory. All that C2 requires is that on one possible way of resolving the context-dependence of counterfactual constructions, they express strict conditionals. And this seems very plausible, especially as it is very natural to phrase strict conditional claims in counterfactual terms.

It turns out that C1 and C2 suffice to pick out a counterfactual conditional as broadest (and uniquely so, up to functional equivalence): according to C1 and C2, the metaphysically strict conditional is broadest. This is in fact trivial, since C2 states that this conditional is a counterfactual conditional, and C1 states that it is at least as broad as every counterfactual conditional. Just as *being* ⊤ can be singled out as the broadest necessity, the metaphysically strict conditional can therefore be singled out as the broadest counterfactual conditional. But for the purposes of doing metaphysics, this is a pyrrhic victory. Counterfactuals are usually appealed to in order to draw distinctions which cannot be drawn using metaphysical necessity alone. The fact that the broadest counterfactual conditional, as identified here, is definable in terms of metaphysical necessity means that these ambitions cannot be satisfied.

In order to escape this negative conclusion, one might reject C1 or C2. One option would be to endorse logical principles for counterfactual conditionals which rule out the metaphysically strict conditional, which would provide grounds for rejecting C2. Although strict conditionals satisfy the most commonly agreed upon logical principles of counterfactuals, there are some principles which have sometimes been endorsed which the (metaphysically) strict conditional does not satisfy. Two prominent examples are the principles of CONDITIONAL EXCLUDED MIDDLE and WEAK CENTERING, both of which are discussed by Lewis (1973):

(CEM) $(p \mathbin{\Box\!\!\rightarrow} q) \vee (p \mathbin{\Box\!\!\rightarrow} \neg q)$

(WC) $(p \wedge q) \rightarrow (p \mathbin{\Box\!\!\rightarrow} q)$

Although both of these principles are contentious, they raise the question whether one might distinguish a non-trivial counterfactual conditional as broadest by replacing C2 with additional logical constraints on counterfactuals, including CEM or WC. This is not the place to investigate this idea in any detail. But there are reasons to doubt that, along these lines, a unique counterfactual conditional can be distinguished as broadest in any straightforward way. This can be illustrated model-theoretically, using the selection function semantics of Stalnaker (1968).

Stalnaker's theory takes the form of a possible worlds model theory. On one way of presenting this model theory, every world is associated with a well-order, which orders worlds accord to how close, or similar, they are to the first world. $\varphi \mathbin{\Box\!\!\rightarrow} \psi$ is interpreted as true in a world just in case ψ is true in the closest world in which φ is true. This model theory validates CEM, and WC can be ensured to be valid by assuming that every world is closest to itself. By extending such models to models of $\mathcal{L}^=$, sufficiently strong constraints on CC might ensure that every counterfactual conditional corresponds to an association of well-orders with worlds. But among such counterfactuals, there is clearly no broadest member, except in the trivial case in which there is only one world. In fact, the broadness relation among such counterfactuals is simply identity.

These model-theoretic considerations are by no means a knock-down argument against the proposal. But they illustrate that it may be very hard to ensure that a broadest counterfactual conditional exists, without allowing it to collapse into the metaphysically strict conditional. In contrast to the case of necessity, it is therefore much harder to see how one could argue that there is a distinguished counterfactual conditional at issue in the context of metaphysics.

6

Actuality, and the Necessity of Distinctness and Possibility

The previous two chapters motivated the theory of CLASSICISM. They argued that on the basis of this theory, metaphysical necessity can be identified as the modality $\Box$ of being the tautologous proposition, i.e., $\lambda p.p = \top$. With this, CLASSICISM turned out to be deductively equivalent to a modal system $\vdash^\Box$ which constitutes a natural extension of (propositionally quantified) S4. This chapter argues that these languages can be extended by an actuality operator @, which leads to a non-conservative extension of the logic of necessity and propositional identity. The resulting strengthening contains the necessity of distinctness, as well as the necessity of possibility (the characteristic principle of S5).

6.1 "Actually"

Modal logicians have found it useful to enrich languages of modal logic with a so-called "actuality" operator. This operator is inspired by certain uses of "actually" in English. Adapting an example from Crossley and Humberstone (1977, p. 12), consider the following sentence:

Possibly, everything which is red is shiny.

Contrast this with the following sentence:

Possibly, everything which is actually red is shiny.

On the intended non-epistemic reading of "possibly," it is natural to take these two sentences to make different claims. Let b be a red ball. A possibility witnessing the truth of the first sentence only requires b to

The Foundations of Modality: From Propositions to Possible Worlds. Peter Fritz, Oxford University Press.
© Peter Fritz 2023. DOI: 10.1093/oso/9780192870025.003.0007

be shiny if it is then still red. In contrast, a possibility witnessing the truth of the second sentence requires b to be shiny, no matter which color it then has. Similarly, consider:

Possibly, everything which is red is actually shiny.

If b is not shiny, then a possibility witnessing the truth of this sentence requires b not to be red: since b is not shiny, b would then not be actually shiny.

Such cases motivate the introduction of a unary operator @ capturing the relevant uses of "actually." On the relevant interpretation, @p could only be true if p is in fact true, in which case @p could not have been false. This can be summed up in the following two principles governing @:

(A1) $\Diamond @p \to p$
(A2) $p \to \Box @p$

Here, possibility is formalized using $\Diamond$, which will be understood as an abbreviation for $\neg\Box\neg$, as is often done in modal logic. Alternatively, given the account of metaphysical necessity of the previous chapter, $\Diamond$ could also be introduced as $\lambda p.p \neq \bot$; doing so would make no difference for present purposes.

In introducing the actuality operator @, it is often acknowledged that some uses of "actually" may not express what @ captures. Crossley and Humberstone (1977) call the use appealed to so far the "logical use," and distinguish it from the "rhetorical use," which they illustrate with the following sentence (p. 11):

Actually it was March, not April, when we bought the house.

It is sometimes quite difficult to distinguish between the logical and rhetorical uses of "actually," partly because many natural modal claims in English involve the use of subjunctive mood, which complicates matters, as discussed by Wehmeier (2004). This is the reason why the examples used above are somewhat stilted. In fact, some recent authors have argued that what Crossley and Humberstone (1977) call the "logical use" cannot be found in ordinary English, but is purely an invention of logicians; see Yalcin (2015), Haraldsen (2015), and Mackay (2017).

For present purposes, the question how the English word "actually" is ordinarily used is ultimately unimportant. What matters is only whether an operator @ can be introduced for which A1 and A2 are true, on any interpretation of p. Extending the methodology of earlier chapters, this seems highly plausible. First, whether it is available in ordinary English or only employed among a small group of modal logicians, there is no reason to doubt that "actually" can be used in accordance with what Crossley and Humberstone call the "logical use." Second, the principles A1 and A2 are consistent with CLASSICISM; in fact, these two principles pin down @ uniquely, as far as this can be expressed in $\mathcal{L}^=$, as will now be shown.

First, it is easy to see that A1 and A2 are respectively equivalent, given CLASSICISM, to the following two principles:

(A1')　$\neg p \rightarrow (@p = \bot)$
(A2')　$p \rightarrow (@p = \top)$

Thus A1 and A2 settle, for every p, the identity of @p as $\top$ or $\bot$, depending on whether p is true or false. This amounts to a complete specification of the applicative behavior of @. No reason has been encountered why it should not be possible to introduce @ along present lines, and the metasemantic basis for @ appears to be especially rich—richer, certainly, than that for the quantifiers of $\mathcal{L}$. In the following, it will therefore be assumed that @ has been successfully introduced.

The following sections explore the effect of adding A1 and A2 to CLASSICISM. Before turning to these issues, two remarks are in order. First, although we have not encountered any reasons to think that the introduction of @ fails, there are various views which one might hold on independent grounds which entail that there is no operator satisfying A1 and A2. One such view will be mentioned in section 10.1. Second, although the interaction of @ and □ as the two connectives are introduced here matches their interaction in existing discussions of actuality operators in modal logic precisely, such discussions sometimes make assumptions about the interaction of □ with other modalities which contradict the theory developed up to now. As a downstream consequence, the theory developed up to now will in some ways also disagree with some of the established discussions concerning the interaction of @ with other modal operators. For example, consider a temporal modality G, roughly standing for "it is always going to be the case." Assuming it is Tuesday, A2 entails that it is necessarily actually Tuesday (paraphrasing $\mathcal{L}$ in English). Thus, that it is actually Tuesday is the tautologous proposition $\top$, which is always going to be the case. So, the account developed here predicts that on the use of "actually" corresponding to @, if it is Tuesday then it is always going to be the case that it is actually Tuesday. In general, the present account will be committed to the principle $p \to G@p$, which some existing logics of actuality reject, such as the one developed by Kaplan (1989 [1977], 1978). For further discussion of these issues, see Dorr and Goodman (2020).

6.2 The Necessity of Distinctness

Let $\mathcal{L}^{=@}$ be $\mathcal{L}^=$ with @ receiving the intended interpretation introduced here. Define C@ to be the extension of CLASSICISM by A1 and A2:

Definition 6.1. C@ $:=$ CLASSICISM $+$ A1 $+$ A2.

By adapting an argument by Williamson (1996), we can derive the necessity of distinctness (of propositions) from C@. This is regimented in $\mathcal{L}^=$ as the following principle:

(ND) $p \neq q \to \Box(p \neq q)$

Proposition 6.2. C@ $\vdash^= $ ND.

Proof. The claim is shown by the following deduction. It makes use of inferences of S4 modal logic; this is justified by the results of the previous chapter:

(1) $p \neq q \to \Box\neg@(p = q)$ — A1
(2) $\Box((p = q) \to (@(p = p) \to @(p = q)))$ — CLASSICISM
(3) $\Box@(p = p)$ — RI, A2
(4) $\Box((p = q) \to @(p = q))$ — 2, 3
(5) $p \neq q \to \Box(p \neq q)$ — 1, 4

$\Box$

ND is an example of the non-conservativeness of C@ over CLASSICISM in the @-free fragment, since ND cannot be derived from CLASSICISM alone. This last claim is established in the supplementary section 6.6.

6.3 The Necessity of Possibility

As the previous section shows, the introduction of @ settles matters left open by CLASSICISM, such as ND. Another matter settled this way is 5, the characteristic principle of the modal logic S5, stating that what is possible is necessarily possible:

(5) $\Diamond p \to \Box \Diamond p$

Indeed, this follows from the fact that ND and 5 are equivalent, relative to CLASSICISM, in the following sense:

Proposition 6.3. CLASSICISM + ND $\vdash^=$ 5 *and* CLASSICISM + 5 $\vdash^=$ ND.

Proof. The first claim is straightforward, as 5 is equivalent to the following instance of ND:

$$p \neq \bot \to \Box(p \neq \bot)$$

For the second claim, note that ND is equivalent to:

$$\neg(p = q) \to \neg(p = q) = \top$$

By Lemma 5.1, CLASSICISM contains:

$$(p = q) = ((p \leftrightarrow q) = \top)$$

So ND follows from the following:

$$(p \leftrightarrow q) \neq \top \to ((p \leftrightarrow q) \neq \top) = \top$$

And this is equivalent to an instance of 5. □

6.4 Strengthening Classicism

So far, an actuality operator @ was introduced, which led to the language $\mathcal{L}^{=@}$. With the stipulative principles governing @, CLASSICISM was strengthened to C@. This was shown to settle two important questions left open by CLASSICISM, namely the necessity of distinctness and possibility, as captured by the principles ND and 5. The resulting logic of necessity and actuality is a (conservative) extension of the standard extension of S5 by an actuality operator, as investigated by Crossley and Humberstone (1977). Supplementary section 6.7 shows that A1 and A2 can also be used to provide an elegant variant axiomatization of this logic.

Having used @ to motivate ND and 5, we will return to working in the language $\mathcal{L}^=$ in the following. Instead of relying on A1 and A2, we will therefore strengthen CLASSICISM directly to ensure that ND and 5 are derivable. Since they are equivalent relative to CLASSICISM, it suffices to add either one of them. Making an arbitrary choice, we will add 5. It does, however, matter *how* this principle is added to the commitments made so far. Recall that Chapter 4 opted for a maximally strong way of incorporating the idea that provably equivalent formulas express the same propositions. This was done by adding the RULE OF EQUIVALENCE to the proof system $\vdash^=$, and endorsing all principles provable in the resulting system. In order to carry this approach over to the additional commitment of 5, we define:

Definition 6.4. C5 *is the set of theorems of* $\vdash^=$ + 5 + RE, *the proof system resulting from adding to* $\vdash^=$ *the axiom* 5 *and rule* RE.

There are several reasons for proceeding with C5 rather than C@. First, C5 is simpler to formulate than C@ in that the former allows us to continue with $\mathcal{L}^=$ rather than $\mathcal{L}^{=@}$. Second, even those who have qualms about the status of actuality may find C5 compelling, on independent grounds. One such independent ground could be provided by an extension of the search for simple and strong theories used to motivate CLASSICISM. The necessity of distinctness and possibility are principles which are widely considered to be attractive in metaphysics. In an informal sense, they are also clearly strong and simple. This provides at least some weak additional motivations for adopting an extension of CLASSICISM which incorporates these principles.

A third and final reason concerns the RULE OF EQUIVALENCE. Whereas C5 is closed under RE by definition, closing the consequences of C@ in $\vdash^=$ under RE brings us back to EXTENSIONALISM. This is easily seen. By A1 and A2, C@ $\vdash^=$ $p \leftrightarrow$ @p. With RE, one would obtain $p =$ @p, and so via A2: $p \to (p = \top)$. Analogously $\neg p \to (p = \bot)$, so it would follow that every proposition is either $\top$ or $\bot$, which entails EXTENSIONALISM. But this conclusion cannot be established from C@ in $\vdash^=$, as can be shown using standard possible worlds models. This derivation of EXTENSIONALISM may be seen as a variant of the so-called *slingshot* arguments of Church (1943) and Gödel (1984 [1944]). But this observation does not provide any strong reasons to doubt the earlier conclusion that EXTENSIONALISM is false. Rather, it merely shows that the logic of actuality cannot be closed under RE. This corresponds to the now standard viewpoint on the logic of actuality, which is that any logic of actuality which includes $p \leftrightarrow$ @p as a theorem cannot be closed under necessitation. Chapter 8 will discuss the addition of a new kind of quantifiers to $\mathcal{L}^=$. Obtaining a natural extension of CLASSICISM to these quantifiers requires appealing to RE. This is only possible in the context of C5 rather than C@, in order to escape the collapse into EXTENSIONALISM. This is the third reason for proceeding with C5.

6.5 The Logic of Necessity Revisited

The previous chapter characterized precisely the logic of necessity corresponding to CLASSICISM. With CLASSICISM now being strengthened to C5, this raises the question how much the corresponding logic of necessity is strengthened. The answer to this question turns out to be straightforward. It suffices to add the standard modal principle 5 as an axiom to $\vdash^\square$. Call the resulting proof system $\vdash_5^\square$. It is routine to show an analog of Proposition 5.6 using C5 instead of CLASSICISM, and the consequences of $\vdash_5^\square$ instead of $\vdash^\square$.

The resulting modal system characterizes exactly the modal principles entailed by C5, but the language $\mathcal{L}^\square$ used in this proof system involves some higher-order quantifiers not usually found in discussions of modal logic. It may therefore be useful to characterize the resulting logic of some fragments of $\mathcal{L}^\square$ in more familiar terms. It turns out that this is also easily done, and that the relevant fragments are very well-known modal logics. Restricting ourselves to formulas in propositionally quantified modal logic leads to the standard propositionally quantified extension of S5,

which is called $S_\Pi 5$ in Fritz (forthcoming). (Bull (1969) calls this logic
S5Π; it also corresponds to the logic S5π in Fine (1970).) This logic
is itself a conservative extension of the standard propositional modal
logic S5.

To make these results precise, we now define the languages of propositional modal logic and propositionally quantified modal logic. As usual
in modal logic, any truth-functionally complete set of Boolean connectives can be used, and only one quantifier is required.

Definition 6.5. *Let $\mathcal{L}_m$ be the language based on propositional variables
$p, q, \ldots$ with formulas built up using Boolean connectives $\neg$ and $\wedge$, and a
necessity operator $\Box$. That is, the formulas of $\mathcal{L}_m$ are recursively specified
as follows:*

(1) *If p is a propositional variable, then p is a formula.*
(2) *If φ and ψ are formulas, then $\neg\varphi$, $\varphi \wedge \psi$, and $\Box\varphi$ are formulas.*

*Let $\mathcal{L}_p$ be the extension of $\mathcal{L}_m$ by propositional quantifiers. That is, the
formulas of $\mathcal{L}_p$ are recursively specified by (1), (2) and:*

(3) *If φ is a formula and p is a propositional variable, then $\forall p\varphi$ is a
   formula.*

In the context of $\mathcal{L}_m$ and $\mathcal{L}_p$, other Boolean connectives are considered
as defined in one of the truth-functionally adequate ways, and existential
quantifiers are treated as duals of universal quantifiers as usual. As
above, $\Diamond$ is treated as the dual of $\Box$. S5 and $S_\Pi 5$ can now be defined as
follows:

Definition 6.6. *Let S5 be set of formulas of $\mathcal{L}_m$ which are derivable in the
proof system $\vdash_{S5}$ consisting of the following axiom schemas and rules:*

(Taut)	*Tautologies*	(MP)	$\varphi, \varphi \rightarrow \psi / \psi$
(K)	$\Box(\varphi \rightarrow \psi) \rightarrow (\Box\varphi \rightarrow \Box\psi)$	(N)	$\varphi/\Box\varphi$
(T)	$\Box\varphi \rightarrow \varphi$	(5)	$\Diamond\varphi \rightarrow \Box\Diamond\varphi$

*Let $S_\Pi 5$ be set of formulas of $\mathcal{L}_p$ which are derivable in the proof system
$\vdash_{S_\Pi 5}$ consisting of the above axiom schemas and rules (in $\mathcal{L}_p$), as well as
the following:*

(UI)	$\forall p\varphi \rightarrow \varphi[\psi/p]$	(CUG)	$\varphi \rightarrow \psi / \varphi \rightarrow \forall p\psi$ *(p not free in φ)*

For the statement of UI, recall that $\varphi[\psi/p]$ is only defined if ψ is free for p in φ. The principles of quantification used in $S_\Pi 5$ are easily seen to be equivalent to those used in $\vdash$. (The reason for deviating from the earlier ones is just simplicity. The more complex ones were used above since they allowed a simpler argument for their correctness.) The claims can now be proven:

Proposition 6.7. *For any $\varphi \in \mathcal{L}_m$ and $\psi \in \mathcal{L}_p$:*
(1) $\vdash_{S5} \varphi$ *iff* $\vdash_5^\square \varphi$ *iff* $C5 \vdash^= \varphi^\ddagger$.
(2) $\vdash_{S_\Pi 5} \psi$ *iff* $\vdash_5^\square \psi$ *iff* $C5 \vdash^= \psi^\ddagger$.

Proof. In each case, the second *iff* is immediate from the extension of Proposition 5.6 mentioned above.

Consider first the remaining claim of (2). That $\vdash_{S_\Pi 5} \psi$ only if $\vdash_5^\square \psi$ follows from an induction on the length of proofs. The argument is straightforward, since every one of the relevant axioms and rules is included in $\vdash_5^\square$. The converse direction follows from the completeness theorem of Holliday (2019): Holliday considers a model theory for propositionally quantified modal logic based on complete Boolean algebras. In such a model, the elements of the algebra serve as propositions. Universally quantified statements are interpreted as the greatest lower bounds of their instances, and attributions of necessity as 1 or 0 (the top and bottom elements of the algebra), depending on whether the complement clause is interpreted as 1 or not. Holliday's completeness result shows that if $\nvdash_{S_\Pi 5} \psi$, then ψ is not valid on every such model. In section 8.6, we will show that these models are straightforwardly extended to serve as models for $\mathcal{L}^=$, and Proposition 8.6 shows that they validate C5. Further, such models interpret a formula ψ of $\mathcal{L}_p$ in the same way as the corresponding formula $\psi^\ddagger$ of $\mathcal{L}^=$. This follows by a routine induction on the complexity of ψ. Therefore, if $\nvdash_{S_\Pi 5} \psi$, then ψ is not valid on all models based on complete Boolean algebras. Consequently, the same holds for $\psi^\ddagger$; so $C5 \nvdash^= \psi^\ddagger$, whence $\nvdash_5^\square \psi$.

The remaining claim of (1) now follows from the fact that $S_\Pi 5$ is a conservative extension of S5. This in turn is straightforward along similar lines to the argument just discussed. That $\vdash_{S5} \varphi$ only if $\vdash_{S_\Pi 5} \varphi$ follows by a trivial induction on the length of proofs. The converse direction follows straightforwardly from the completeness of S5 with respect to possible worlds models in which $\Box\varphi$ is true in a world just in case φ is true in every world. If $\nvdash_{S5} \varphi$, then it is false in some such possible worlds model. Since this serves also to interpret $\mathcal{L}_p$, and $S_\Pi 5$ is sound with respect to such models, $\nvdash_{S_\Pi 5} \varphi$. $\Box$

This result shows that although the systems discussed here in various extensions of $\mathcal{L}_m$ are unfamiliar in that they incorporate various higher-order resources, the logic of necessity which falls out of the theory C5 motivated here is very familiar, and well-understood. For example, by a well-known arguments using the B axiom, which is derivable in S5, one can derive the Barcan formula, i.e., the instances of the following schema:

(BF) $\forall p \Box \varphi \rightarrow \Box \forall p \varphi$

The argument goes back to Prior (1956); for an exposition, see Hughes and Cresswell (1996, p. 247).

6.6 Supplement: Independence

This supplementary section shows that CLASSICISM does not entail the necessity of distinctness (ND), and so also not the necessity of possibility (5). This will be done using a single model, which is chosen to be as simple as possible. It can be understood to correspond to a very simple possible worlds model of the failure of the 5 axiom in the context of propositional modal logic, using the accessibility relations of Kripke (1963): a model with two worlds w and v, where v is accessible from w but not vice versa, and each world is accessible from itself. Transposing this to the setting of $\mathcal{L}^=$, where identity rather than necessity is primitive, the model counts propositions as identical in w if they agree in truth-value in both worlds (both being accessible from w), and identical in v if they agree in truth-value in v (being the only world accessible from v). This basic model-theoretic idea requires some refinement in order to accommodate the higher-order quantifiers of $\mathcal{L}$. This can be done using techniques employed for the same purpose, but in much greater generality, in Bacon (2018a).

Let $\mathfrak{M} = \langle W, D, V \rangle$ be a possible worlds model satisfying the following desiderata. First, W is a two-element set $\{w, v\}$ and D the constant but restricted domain function such that for all $n > 0$ and $x \in W$:

$$D_0^x := P_0$$
$$D_n^x := \{o \in P_n : \text{if } d_1 \sim e_1, \ldots, d_n \sim e_n \text{ then } o(\bar{d}) \sim o(\bar{e})\}$$

where and $d \sim e$ is defined as: $v \in d$ iff $v \in e$.

The relation $\sim$ is extended to elements of P_n, and further to assignment functions, by letting $o_1 \sim o_2$ iff in general $d_1 \sim e_1, \ldots, d_n \sim e_n$ only if $o_1(\bar{d}) \sim o_2(\bar{e})$, and $a \sim b$ iff $a(x) \sim b(x)$ for all variables x. Second, $V(=)$ is such that for all $d, e \in P_0$:

$$V(=)(\langle d, e \rangle) = \begin{cases} W & \text{if } d = e \\ \varnothing & \text{if } d \nsim e \\ \{v\} & \text{otherwise} \end{cases}$$

Lemma 6.8. *For every* $x \in W$ *and* x-*assignment* a, $[\![\varepsilon]\!]^a \in D_n^x$.

Proof. Only the case of λ-terms is interesting. This can be shown using the following claim, which can be established by induction on the complexity of ε: if $a \sim b$ then $[\![\varepsilon]\!]^a \sim [\![\varepsilon]\!]^b$. $\square$

Lemma 6.9. *Every theorem of* $\vdash^=$ *is valid on* $\mathfrak{M}$.

Proof. By induction on the length of proofs. The case of UI follows by Lemmas 2.9 (2) and 6.8. The w-validity of LL is immediate; its v-validity follows from the definition of $\sim$. The validity of RI is immediate. The remaining cases follow by Lemma 2.10. $\square$

Lemma 6.10. CLASSICISM *is valid on* $\mathfrak{M}$.

Proof. By the proof of Lemma 6.9, it suffices to show that validity on $\mathfrak{M}$ is closed under RE, which is immediate. $\square$

Proposition 6.11. CLASSICISM $\nvdash^=$ ND.

Proof. With Lemma 6.9, it suffices to show that ND is not valid on $\mathfrak{M}$. This is easily seen, letting $a(p) = \varnothing$ and $a(q) = \{w\}$: In this case, $w \in [\![p \neq q]\!]_a$. But $v \notin [\![p \neq q]\!]_a$, whence $w \notin [\![(p \neq q) = \top]\!]_a$.   □

6.7 Supplement: Axioms for Actuality

Axioms A1 and A2 can be used to provide an axiomatization of the standard propositional modal logic of necessity and actuality which is arguably simpler and more natural than the most well-known syntactic characterizations. This supplementary section presents this axiomatization, and shows that it captures exactly this logic. (For a related approach to the axiomatics of actuality in the context of conditional logics, see Williamson (2009).)

In their classic discussion, Crossley and Humberstone (1977) operate with a propositional modal logic with two modal operators, □ and @. Extending notation introduced in this chapter, this can be defined as the language $\mathcal{L}_m^@$ which is obtained by extending $\mathcal{L}_m$ with the unary sentential operator @. Crossley and Humberstone first introduce very simple possible world models for this language. A model is determined by a set of worlds W and a function determining which propositional variables are true in which worlds, as well as a distinguished actual world w^*. Truth of a complex formula relative to a world is defined in the usual recursive way, with □φ being true in a world just in case φ is true in every world. Crucially, @φ is then stipulated to be true in a world just in case φ is true in the distinguished actual world w^*.

Using this model theory, Crossley and Humberstone define two notions of validity. A formula is defined to be *generally valid* if it is true in every world of every model. A formula is defined to be *real-world valid* if it is true in the actual world of every model. By definition, it is immediate that the logic of general validity is closed under necessitation: if φ is generally valid, then so is □φ. This is not the case for real-world validity: $p \rightarrow$ @p is real-world valid, but □$(p \rightarrow$ @$p)$ is not real-world valid, as p may be true in some world without being true in the actual world.

Axiomatically, Crossley and Humberstone focus on general validity. They show that the following extension of $\vdash_{S5}$ is sound and complete with respect to general validity (where $\vdash_{S5}$ is now taken to be a proof system with instances of axiom schemas taken from—and rules operating on—formulas of the expanded language $\mathcal{L}_m^@$):

Definition 6.12. *Let* $\vdash_{S5A}$ *be the extension of the proof system* $\vdash_{S5}$ *by the following axiom schemas:*

(CH1) $@(@\varphi \to \varphi)$ (CH2) $@(\varphi \to \psi) \to (@\varphi \to @\psi)$
(CH3) $@\varphi \to \neg@\neg\varphi$ (CH4) $\Box\varphi \to @\varphi$
(CH5) $@\varphi \to \Box@\varphi$

Crossley and Humberstone (1977, p. 15) note that a syntactic characterization of the real-world valid formulas can be derived from this axiomatization by noting that φ is real-world valid just in case $@\varphi$ is generally valid. Equivalently, one can characterize the real-world valid formulas as those deducible from the generally valid ones and the instances of the schema $\varphi \leftrightarrow @\varphi$ using MP.

For many purposes, these syntactic characterizations of real-world validity are perfectly adequate. But it is also interesting to devise a more standard proof system for real-world validity. The difficulty in doing so is that standard proof systems for S5 involve the rule of necessitation, which does not preserve real-world validity. This difficulty can be overcome by starting from a proof system for S5 which does not involve the rule of necessitation. Such a proof system is described by Williamson (2013, p. 110). It turns out that adding (schematic versions of) axioms A1 and A2 for @ produces a proof system sound and complete for real-world validity. This can be defined as follows:

Definition 6.13. *Let* $\vdash_{S5@}$ *be the proof system in* $\mathcal{L}_m^@$ *consisting of the following axiom schemas and rules:*

($\Box$TAUT) $\Box\varphi$ (φ a tautology)   (MP) $\varphi, \varphi \to \psi / \psi$
(K) $\Box(\varphi \to \psi) \to (\Box\varphi \to \Box\psi)$ (T) $\Box\varphi \to \varphi$
(4) $\Box\varphi \to \Box\Box\varphi$ (5') $\neg\Box\varphi \to \Box\neg\Box\varphi$
(A1$_S$) $\Diamond@\varphi \to \varphi$ (A2$_S$) $\varphi \to \Box@\varphi$

Given the above-mentioned result that φ is real-world valid iff $@\varphi$ is deducible in the proof system of Crossley and Humberstone, it suffices to establish the following proposition:

Proposition 6.14. *For every formula φ of $\mathcal{L}_m^{@}$: $\vdash_{S5@} \varphi$ iff $\vdash_{S5A} @\varphi$.*

Proof. It is routine to show that $\vdash_{S5@}$ is sound with respect to real-world validity, from which it follows that $\vdash_{S5@} \varphi$ only if φ is real-world valid. By the completeness facts noted above, φ is real-world valid only if $\vdash_{S5A} @\varphi$. This establishes the left to right direction.

For the remaining direction, we first establish the following claim by an induction on the length of proofs in $\vdash_{S5A}$:

(∗)   If $\vdash_{S5A} \varphi$ then $\vdash_{S5@} \Box\varphi$.

The case of Taut is immediate, by □Taut. The cases of K, T, and 5 can be proven along the lines of the argument in Williamson (2013, p. 110, fn. 36). For the rule of N of necessitation, assume that $\vdash_{S5A} \varphi$ whence $\vdash_{S5A} \Box\varphi$ by N. Then by induction hypothesis, $\vdash_{S5@} \Box\varphi$, whence $\vdash_{S5@} \Box\Box\varphi$ using 4. The case of MP is similar, and straightforward.

It remains to derive the necessitations of the five axioms involving @. Given the induction cases demonstrated so far, it can be assumed that $\vdash_{S5@}$ proves every theorem of S5. The following argument establishes (CH1):

(1) $@\varphi \to \varphi$ A1$_S$, T
(2) $\Box@(@\varphi \to \varphi)$ 1, A2$_S$

The remaining cases can all be demonstrated using a case-distinction, distinguishing the different truth-values the relevant formula(s) can take. Exemplarily, consider the case of CH5:

(1) $\neg\varphi \to \Box\neg@\varphi$ A1$_S$
(2) $\varphi \to \Box\Box@\varphi$ A2$_S$, 4
(3) $\Box(@\varphi \to \Box@\varphi)$ 1, 2

Having established ∗, the remaining right to left direction follows easily: By A1$_S$ and T, $\vdash_{S5@} \Box@\varphi \to \varphi$. So if $\vdash_{S5A} @\varphi$, then $\vdash_{S5@} \Box@\varphi$ by ∗, whence $\vdash_{S5@} \varphi$ with the claim just noted.   □

PART IV
WORLDS

7

A Theory of Possible Worlds

7.1 Worlds as Propositions

The theory C5 arrived at in the previous chapter is a relatively orthodox modal theory of propositional individuation. A notion of necessity is identified which obeys the strong modal logic S5, and propositions turn out to be identical just in case they are necessarily equivalent. This theory can be modeled by possible worlds models in which necessity amounts to truth in all possible worlds. Propositions are then identical if they are true in the same possible worlds; consequently, propositions can be identified with sets of possible worlds.

In the context of set-theoretic models, such talk of possible worlds is unproblematic. In this context, models are mathematical structures in which the elements of a set W are called "possible worlds." Despite this terminology, W may in fact be any set, such as the set of natural numbers. But in many discussions in philosophy which are not purely formal, appeals to possible worlds are more substantial. For example, semantic theories which explain the meaning of a sentence in terms of the possible worlds in which it is true, such as those in the tradition of Montague (1974), cannot just take possible worlds to be arbitrary things. As Barwise and Perry (1985, p. 116) put the point:

> The Montague Grammarian, or other possible worlds theorist, is committed to possible worlds and needs to tell us what they are if we are to take their theory seriously [. . .]. Saying that they are "just indices" is not a responsible response.

The idea that there really is a special class of things which are the possible world has been challenged. As noted by Stalnaker (1976), various authors including Mackie (1973) and Powers (1976) simply balk at the very idea; see also Kearns (1981, p. 86). Others have pointed to

The Foundations of Modality: From Propositions to Possible Worlds. Peter Fritz, Oxford University Press.
© Peter Fritz 2023. DOI: 10.1093/oso/9780192870025.003.0008

particular features which possible worlds are meant to have as being problematic. A common point of contention is the idea that possible worlds are maximal, in the sense of being fully specific alternatives. For example, Hale (2013, pp. 230–231) and Rumfitt (2015, §6.2) advocate theorizing in terms of (possibly) incomplete *possibilities* rather than possible worlds. Similar proposals are explored by Humberstone (1981); Barwise and Perry (1985); Fine (2017c) and Holliday (2021). Indeed, even authors who often employ possible worlds in their theorizing, and who have contributed substantially to the widespread use of the notion in contemporary philosophy, express sympathy with this viewpoint; see Kripke (1980 [1972], pp. 15–20) and Stalnaker (2012, pp. 12–13).

In response, various accounts of possible worlds have been proposed which identify them with various kinds of entities perceived to be less problematic. For example, Lewis (1986)—see also Lewis (1968, 1973)—suggests that possible worlds are maximal connected space-time regions. Fine (2005 [1977]), developing ideas by Prior (1967, ch. V), proposes to consider worlds as certain propositions, as do Daniels and Freeman (1977) and Reinhardt (1980). Similarly, possible worlds have been suggested to be certain states of affairs (Plantinga 1974, 1976); certain sets of propositions (Adams 1974, 1981); and certain properties of the world (or: ways a world might be) (Forrest 1986; Stalnaker 1976, 2003, 2012).

The extent to which the different options have been developed varies widely. As noted by Menzel (2016, "Supplement on Problems with Abstractionism"), some have been put forward with little formal rigor, and in certain cases, the proposed account of possible worlds even turns out to be inconsistent. Furthermore, most authors do not articulate clearly the requirements which a reductive account of possible worlds would have to satisfy in order to be successful.

In this chapter, we will approach the issue of possible worlds from an abstract perspective. Based on the theory of necessity and propositional identity developed in the previous chapters, we begin by formulating a general theory of possible worlds. This theory will later turn out to be complete, in the sense of settling every question about possible worlds which can be formulated in the relevant language. We will see that if anything satisfies this theory—i.e., if there is any way of making sense of talk of possible worlds—then certain 'maximal' propositions satisfy this theory as well. Questions about the nature of possible worlds may therefore be set aside, as possible worlds may be taken to be certain maximal propositions without precluding any options. The question whether there are possible worlds can thus be reduced to the question whether maximal propositions play the role of possible worlds. This can in turn be reduced to a single axiomatic principle of ATOMICITY. The question whether talk of possible worlds is in good standing therefore amounts to the question whether ATOMICITY is true. We will take up this question in the next chapter. For brevity, talk of possible worlds will from now on be shortened to talk of just "worlds."

7.2 Formalizing World-Talk

In order to investigate the notion of a world without any preconception about what worlds might be, it will be useful to introduce to the languages used above world quantifiers whose interpretation is left open. In fact, not all of the resources of $\mathcal{L}$ will be required; in particular, there will be no need for quantification over modalities. Further, the discussion of worlds is most naturally couched in a setting in which necessity rather than identity is taken as primitive. We therefore add world quantifiers to the propositionally quantified modal language $\mathcal{L}_p$.

So, let $w, v, \ldots$ be a countable set of new "world" variables, which may be bound by a universal quantifier $\forall$. To talk about identity of worlds, we include a connective $=$ which may be flanked by world variables. To talk about the truth of a proposition in a world, we include a connective $\triangleright$ which takes as arguments a world variable w (on the left) and a formula φ (on the right) to form a formula $w \triangleright \varphi$ stating that φ is true in w. We add these to $\mathcal{L}_p$:

Definition 7.1. *Let $\mathcal{L}_w$ be the language which adds to the recursive clauses 1–3 of $\mathcal{L}_p$ the following clauses, using world variables $w, v, \ldots$:*
  (4) *If φ is a formula and w is a world variable, then $\forall w\varphi$ and $w \triangleright \varphi$ are formulas.*
  (5) *If w and v are world variables, then $w = v$ is a formula.*

p is said to *strictly imply* q if necessarily, if p then q. Following Lewis and Langford (1959 [1932]), we will abbreviate this as follows:

$$\varphi \dashv 3 \, \psi := \Box(\varphi \rightarrow \psi)$$

Next, we specify a proof system for $\mathcal{L}_w$. Since C5 has been shown to lead to all the theorems of $S_{\Pi}5$, we take the standard axiomatization of $S_{\Pi}5$ stated in the previous chapter as a starting point, and extend the usual principles of quantification and identity for propositions to worlds. As in the propositional case, these principles for worlds lead to consequences which may be controversial. For example, the resulting system proves necessitism about worlds, i.e., the claim that it is a necessary matter which worlds there are. However, worlds are plausibly sufficiently similar to propositions that necessitism about propositions makes necessitism about worlds a natural starting assumption. Furthermore, we will ultimately see that the theoretical role of worlds as articulated on the basis of the principles in this proof system is at least not unduly restrictive, since the next chapter will argue for ATOMICITY and therefore the claim that maximal propositions play the role of worlds.

Definition 7.2. *Let $\vdash_w$ be the proof system extending $\vdash_{S_{\Pi}5}$ by the following axiom schemas and rule:*

(UI^w) $\forall w\varphi \rightarrow \varphi[v/w]$ (CUG^w) $\varphi \rightarrow \psi/\varphi \rightarrow \forall w\psi$ *(w not free in φ)*
(RI^w) $w = w$ (LL^w) $w = v \rightarrow (\varphi(w) \rightarrow \varphi(v))$

As in the case of $\vdash$, we will use $\Gamma \vdash_w \varphi$ to state the existence of $\gamma_1, \dots, \gamma_n$ such that $\vdash_w \bar{\forall}\gamma_1 \wedge \dots \wedge \bar{\forall}\gamma_n \rightarrow \bar{\forall}\varphi$. It is useful to note that in cases in which the premises entail their necessitations, a deduction theorem can be proven, as the following lemma shows. From the details of the proof, it is immediate that the analogous claim holds for $\vdash_{S_{\Pi}5}$ as well.

Lemma 7.3. *If $\Gamma \vdash_w \Box\gamma$ for all $\gamma \in \Gamma$, then $\Gamma \vdash_w \varphi$ iff $\vdash_w^\Gamma \varphi$, where $\vdash_w^\Gamma$ is the proof system obtained by adding the elements of Γ as axioms to $\vdash_w$.*

Proof. That $\Gamma \vdash_w \varphi$ only if $\vdash_w^\Gamma \varphi$ is immediate. For the converse direction, we show by an induction on the length of proofs that if $\vdash_w^\Gamma \varphi$, then:

$$\{\Box\gamma : \gamma \in \Gamma\} \vdash_w \Box\varphi$$

The crucial case of the rule of necessitation follows using the fact that $4 = \Box p \rightarrow \Box\Box p$ is a theorem of S5. With the assumption that $\Gamma \vdash_w \Box\gamma$

for all $\gamma \in \Gamma$, it follows that $\vdash_w^\Gamma \varphi$ only if $\Gamma \vdash_w \Box\varphi$, whence $\Gamma \vdash_w \varphi$ using the fact that $T = \Box p \to p$ is an axiom of $\vdash^w$. □

Finally, recall that $\mathcal{L}_p$, and so $\mathcal{L}_w$, does not contain a primitive connective for propositional identity. However, propositional identity has been argued to amount to necessary equivalence, which motivates the following definition in the context of $\mathcal{L}_p$ and $\mathcal{L}_w$:

$$\varphi = \psi := \Box(\varphi \leftrightarrow \psi)$$

Along the lines of Proposition 5.6, it is routine to show that the standard axioms of the reflexivity of identity and Leibniz's Law can be derived in $\vdash_w$ using this definition.

7.3 A Theory of Worlds

$\vdash_w$ imposes minimal constraints on worlds: it merely requires quantification over worlds and identity among worlds to obey basic logical principles. The next task is therefore to develop a more substantial theory of worlds, articulating the theoretical role worlds are meant to play in theorizing about modality. Two central principles are obvious; they connect necessity and possibility to truth in all and some (possible) worlds:

(WN) $\Box p \leftrightarrow \forall w(w \triangleright p)$
(WP) $\Diamond p \leftrightarrow \exists w(w \triangleright p)$

Relative to $\vdash_w$, it turns out that these two principles need to be augmented by a number of auxiliary principles. As section 7.6 will show more rigorously, WN and WP allow for situations which clearly contradict the theoretical role of worlds. For example, these two principles are compatible with there being exactly two worlds, one in which just the necessary propositions are true, and one in which just the possible propositions are true, despite there being a contingent proposition p. In this scenario, neither p nor $\neg p$ is true in the first world, whereas both are true in the second world. A natural way of ruling out such deviant worlds is to require that every world could accurately describe how things are. Call this *being actual*; it can be defined as follows:

$$\mathcal{A}w := \forall p(p \leftrightarrow w \rhd p)$$

These problematic cases can then be ruled out by imposing an auxiliary principle according to which every world could be actual.

Another kind of deviant behavior of worlds which is not ruled out by WN and WP makes it contingent which propositions are true in which worlds. It may of course be contingent which propositions are true (*simpliciter*), and it may be contingent which world is actual. But it would at least be odd for it to be contingent whether a given proposition is true *in* a given world. This motivates a second auxiliary principle according to which a proposition possibly true in a world is necessarily true in it. Finally, WN and WP do not rule out duplicate worlds, i.e., distinct worlds in which the same propositions are true. But this is also problematic: if worlds w and v are distinct, then the proposition that w is actual should be true in w but not true in v. This motivates a third auxiliary principle according to which worlds are identical if the same propositions are true in them.

In $\mathcal{L}_w$, the three auxiliary principles can be stated as follows. (For much of the following, the universal quantifiers binding world variables in W1 and W2 could be omitted. But including them simplifies the statement and proof of Proposition 7.7 below.)

(W1) $\forall w \Diamond \mathcal{A}w$

(W2) $\forall w(\Diamond w \rhd p \to \Box w \rhd p)$

(W3) $\forall w \forall v(\forall p(w \rhd p \leftrightarrow v \rhd p) \to w = v)$

Using models, it can be shown that none of these three auxiliary principles is derivable from WN and WP. The details are given in a supplementary section 7.6.

A satisfactory theory of worlds should therefore entail all of WN, WP, and W1–3. Is there anything else it should entail? Below, we will see that these five principles suffice, and so that such a theory is complete in an informal sense. Showing this will involve some technical arguments, as well as the principle of ATOMICITY, which will only be argued for in the next chapter. We therefore postpone the discussion of this issue to a supplementary section in the next chapter, section 8.7. For now, we will assume that WN, WP, and W1–3 are informally complete. (No argumentative circularity arises, however, since the argument for ATOMICITY does not depend in any way on this assumption, or more generally on the development in this chapter.)

A second question is whether WN, WP, and W1–3 are independent. The answer to this question is negative. In the presence of W1–3, each of WN and WP can be derived from the other. Indeed, both can be derived from one direction of WP. (Similarly, both can be derived from one direction of WN, but the following formulation will be most useful.) The relevant direction of WP is the following:

(W0) $\Diamond p \rightarrow \exists w(w \rhd p)$

However, adding W0 to W1–3 produces a set of independent principles from which both WN and WP follow, as we will now see. Call this theory W:

$$W := \{W0, W1, W2, W3\}$$

Proposition 7.4. $W \vdash_w$ WN *and* $W \vdash_w$ WP.

Proof. Here and in the following, derivations are not given in full detail, and only the most important principles involved in the relevant steps are specified. Principles of propositional reasoning, including those of S5 propositional modal logic will be used throughout. In the derivations, the members of W can be taken to be axioms; this follows with Lemma 7.3 from the fact that W entails the necessitation of each of its members. The first derivation shows that using W, the converse direction of W0 can be derived; this establishes WP:

(1) $\mathcal{A}w \rightarrow (w \rhd p \rightarrow p)$ UI
(2) $\Box(\mathcal{A}w \rightarrow (w \rhd p \rightarrow p))$ 1, N
(3) $w \rhd p \rightarrow \Box w \rhd p$ W2
(4) $w \rhd p \rightarrow \Box(\mathcal{A}w \rightarrow p)$ 2, 3
(5) $w \rhd p \rightarrow \Diamond p$ 4, W1
(6) $\exists w(w \rhd p) \rightarrow \Diamond p$ 5, UG

From this result, the derivability of WN follows:

$$(1) \quad \Diamond \neg p \leftrightarrow \exists w(w \rhd \neg p) \qquad \text{WP}$$
$$(2) \quad \mathcal{A}w \rightarrow (w \rhd \neg p \leftrightarrow \neg w \rhd p) \qquad \text{UI}$$
$$(3) \quad \Diamond(w \rhd \neg p \leftrightarrow \neg w \rhd p) \qquad 2, \text{W1}$$
$$(4) \quad w \rhd \neg p \leftrightarrow \neg w \rhd p \qquad 3, \text{W2}$$
$$(5) \quad \Diamond \neg p \leftrightarrow \exists w(\neg w \rhd p) \qquad 1, 4$$
$$(6) \quad \Box p \leftrightarrow \forall w(w \rhd p) \qquad 5 \qquad \qquad \Box$$

It can also be shown that W is independent in $\vdash_w$, as none of its members can be derived from the remaining members of W. The proof of this involves an extension of possible worlds models, and so is carried out in supplementary section 7.6.

Before turning to the central question of whether there are any entities playing the role of worlds as articulated by W, it is worth noting a potentially unfamiliar consequence of W in conjunction with the fact that being necessary implies being always the case. To illustrate it, consider a temporally contingent fact p, and a world w. By W1, it is either possible that p is true in w, or possible that $\neg p$ is true in w. By W2, it follows that it is either necessary that p is true in w, or necessary that $\neg p$ is true in w. What is necessary is always true, so p is either always true in w, or $\neg p$ is always true in w. Consequently, worlds should not be thought of as temporally extended, including a whole history of how things might have developed through time; instead, they should be thought of as temporally specific, and as determining how things could have been at a single point of time. In some ways, they might therefore be closer to what is sometimes discussed under the terminology of "centered worlds," e.g., by Quine (1969) and Lewis (1979).

7.4 Maximal Propositions

Are there any things which satisfy the theory of worlds W? This section gives a conditional answer: if W is satisfied by any things, then it is (also) satisfied by maximal propositions. A proposition is *maximal* if it settles every matter: for every proposition, a maximal proposition will determine whether it is true or false. In the present setting, maximality can be articulated as strictly implying every proposition or its negation, but not both.

Formally, the notion of a maximal proposition can therefore be defined as follows, where p is the first variable not free in φ (according to some fixed ordering):

$$M\varphi := \forall p((\varphi \rightarrowtail \neg p) \leftrightarrow \neg(\varphi \rightarrowtail p))$$

Alternatively, a proposition can be defined to be maximal if it is possible and strictly implies, for each proposition, either it or its negation. This is the definition adopted by Gallin (1975, p. 85); see also Fine (1970, p. 339). It will be useful to establish that the two definitions are equivalent, given the modal logic assumed here.

Proposition 7.5. $\vdash_{S_{\Pi}5} Mq \leftrightarrow \Diamond q \wedge \forall p(q \rightarrowtail p \vee q \rightarrowtail \neg p)$.

Proof. By a deduction:

(1)	$((q \rightarrowtail \neg p) \leftrightarrow \neg(q \rightarrowtail p)) \rightarrow (q \rightarrowtail p) \vee (q \rightarrowtail \neg p)$	TAUT
(2)	$Mq \rightarrow \forall p((q \rightarrowtail p) \vee (q \rightarrowtail \neg p))$	1
(3)	$\Box \neg q \rightarrow (q \rightarrowtail \neg p) \wedge (q \rightarrowtail p)$	S5
(4)	$\Box \neg q \rightarrow \neg Mq$	3
(5)	$Mq \rightarrow \Diamond q \wedge \forall p((q \rightarrowtail p) \vee (q \rightarrowtail \neg p))$	2, 4
(6)	$\Diamond q \rightarrow ((q \rightarrowtail p) \vee (q \rightarrowtail \neg p) \rightarrow ((q \rightarrowtail \neg p) \leftrightarrow \neg(q \rightarrowtail p)))$	S5
(7)	$\Diamond q \wedge \forall p((q \rightarrowtail p) \vee (q \rightarrowtail \neg p)) \rightarrow Mq$	6
(8)	$Mq \leftrightarrow \Diamond q \wedge \forall p((q \rightarrowtail p) \vee (q \rightarrowtail \neg p))$	5, 7 $\square$

Due to the principles of S5, modal statements are non-contingent, including statements about strict implication. As remarked above, the Barcan formula and its converse, BF and CBF, are derivable in $\vdash_w$ as well. With these observations, it is routine to show that the following principle, which states that the maximality of propositions is a non-contingent matter, is derivable in $\vdash_w$:

(RM) $\Diamond Mp \rightarrow \Box Mp$

We can now show world-talk to be eliminable in favor of talk of maximal propositions, in the following sense. Instead of quantifying over worlds, we quantify over maximal propositions. Instead of stating the identity of worlds, we state the identity of the relevant maximal propositions. Instead of stating that a proposition is true in a world, we state that it is strictly implied by the relevant maximal proposition. This elimination is successful in the sense that, given W, any statement about worlds is equivalent to—and indeed expresses the same proposition as— the corresponding statement about maximal propositions. To establish this formally, let $\mathcal{L}_w^+$ be the expansion of $\mathcal{L}_w$ by a distinct new proposition variable p_w for every world variable w. (The use of new propositional variables is inessential, but simplifies the presentation.) Let $\cdot^\S$ be the recursive mapping from $\mathcal{L}_w$ to (the world-free fragment of) $\mathcal{L}_w^+$ with the only non-trivial clauses as follows:

$$(\forall w \varphi)^\S := \forall p_w (\mathrm{M}p_w \to \varphi^\S)$$
$$(w = v)^\S := p_w = p_v$$
$$(w \rhd \varphi)^\S := p_w \dashv3 \varphi^\S$$

Theorem 7.6. *For every $\mathcal{L}_w$-sentence φ, W $\vdash_w \varphi^\S = \varphi$.*

The proof of this result involves some lengthier deductions, and is therefore relegated to a supplementary section. It proceeds by two lemmas: Lemma 7.13 shows that given W, for every world w, the proposition $\mathcal{A}w$ that w is actual is a maximal proposition. Moreover, this lemma shows that $\mathcal{A}$ determines a one-to-one correspondence from worlds to maximal propositions. With this, Lemma 7.14 shows that every claim about worlds is equivalent to a claim about the corresponding maximal propositions. In effect, Lemma 7.14 is a generalization of Theorem 7.6 to formulas with free variables, so the latter follows immediately.

Theorem 7.6 shows that if there are any entities playing the role of worlds, and so verifying the principles of the theory of worlds W, then every claim φ about worlds in the language $\mathcal{L}_w$ is equivalent to the corresponding claim $\varphi^\S$ about maximal propositions in the world-free fragment. This applies in particular to the four principles of W themselves. Since they are trivially entailed by W, it follows that W entails that maximal propositions satisfy the four principles governing worlds, in the sense that $\mathrm{W}n^\S$ is derivable from W in $\vdash_w$, for every $n \le 3$. Thus, if any entities satisfy the theory of worlds W, then maximal propositions do so as well. The crucial question is therefore whether maximal propositions satisfy the theory W.

7.5 Atomicity

It turns out to be straightforward to argue that maximal propositions satisfy the three auxiliary principles W1–3 of W, as their images under $\cdot^\S$ are theorems of $S_\Pi 5$.

Proposition 7.7. *W1§, W2§, and W3§ are theorems of $S_\Pi 5$.*

Proof. The following deductions in $\vdash_{S_\Pi 5}$ establish the three claims in turn.

W1§ is $\forall p_w(Mp_w \to \Diamond \forall p(p \leftrightarrow (p_w \mathbin{-\!3} p)))$:

- (1) $Mp_w \to \Diamond p_w$ Prop. 7.5
- (2) $\Box(p_w \to \forall p((p_w \mathbin{-\!3} p) \to p))$ S5
- (3) $Mp_w \to (p_w \to \forall p(\neg(p_w \mathbin{-\!3} p) \to \neg p))$ UI, UG
- (4) $Mp_w \to \Box(p_w \to \forall p(p \to (p_w \mathbin{-\!3} p)))$ 3, RM
- (5) $Mp_w \to \Diamond \forall p(p \leftrightarrow (p_w \mathbin{-\!3} p))$ 1, 2, 4
- (6) W1§ 4, UG

W2§ is $\forall p_w(Mp_w \to (\Diamond(p_w \mathbin{-\!3} p) \to \Box(p_w \mathbin{-\!3} p)))$:

- (1) $\Diamond(p_w \mathbin{-\!3} p) \to \Box(p_w \mathbin{-\!3} p)$ S5
- (2) W2§ 1, UG

W3§ is $\forall p_w(Mp_w \to \forall p_v(Mp_v \to (\forall p((p_w \mathbin{-\!3} p) \leftrightarrow (p_v \mathbin{-\!3} p)) \to p_w = p_v)))$:

- (1) $\forall p((p_w \mathbin{-\!3} p) \leftrightarrow (p_v \mathbin{-\!3} p)) \to ((p_w \mathbin{-\!3} p_v) \wedge (p_v \mathbin{-\!3} p_w))$ UI, S5
- (2) $((p_w \mathbin{-\!3} p_v) \wedge (p_v \mathbin{-\!3} p_w)) \to p_w = p_v$ S5
- (3) W3§ 1, 2, UG

$\square$

This leaves the case of W0§. Modulo some truth-functional equivalences and alphabetic variation of variables, this can be written as follows:

$$\Diamond p \to \exists q(Mq \wedge (q \mathbin{-\!3} p))$$

It will be useful to show that this is inter-derivable with the following claim of ATOMICITY, according to which necessarily, there is a true maximal proposition:

(Atomicity) $\Box \exists p(Mp \wedge p)$

For similar results in a in fuller higher-order setting, see Gallin (1975, p. 85) and Bacon and Dorr (forthcoming, §2.3).

Proposition 7.8. $W0^\S \vdash_{S_\Pi 5}$ Atomicity *and* Atomicity $\vdash_{S_\Pi 5} W0^\S$.

Proof. By two deductions, using the analog of Lemma 7.3 for $\vdash_{S_\Pi 5}$:

$W0^\S \vdash_{S_\Pi 5}$ Atomicity:

(1) $\Diamond \neg \exists p(Mp \wedge p) \rightarrow \exists q(Mq \wedge (q \mathbin{\text{-}\!3} \neg \exists p(Mp \wedge p)))$ $W0^\S$
(2) $Mq \rightarrow \Diamond q$ Prop. 7.5
(3) $(q \mathbin{\text{-}\!3} \neg \exists p(Mp \wedge p)) \rightarrow (q \mathbin{\text{-}\!3} (Mq \rightarrow \neg q))$ UI
(4) $Mq \wedge (q \mathbin{\text{-}\!3} \neg \exists p(Mp \wedge p)) \rightarrow \bot$ 2, 3, RM
(5) Atomicity   1, 4

Atomicity $\vdash_{S_\Pi 5} W0^\S$:

(1) $p \rightarrow \exists q(Mq \wedge q \wedge p)$ Atomicity
(2) $\Diamond p \rightarrow \exists q \Diamond (Mq \wedge q \wedge p)$ 1, BF
(3) $\Diamond (Mq \wedge q \wedge p) \rightarrow \Diamond (Mq \wedge (q \mathbin{\text{-}\!3} p))$ UI
(4) $\Diamond (Mq \wedge (q \mathbin{\text{-}\!3} p)) \rightarrow Mq \wedge (q \mathbin{\text{-}\!3} p)$ RM
(5) $W0^\S$ 2–4

$\square$

The upshot of this chapter is therefore: whether or not talk of worlds can be taken at face value simply depends on whether Atomicity is true. Assessing this claim is the topic of the next chapter.

7.6 Supplement: Consistency Proofs Using Models

This supplementary section defines a class of models in order to prove facts about what cannot be derived in $\vdash_w$. These models provide an algebra A of propositions for propositional quantifiers to range over, as well as a domain D of worlds for world quantifiers to range over. A very general class of such models could be defined by allowing A to be any complete Boolean algebra (see the constructions discussed in section 8.6), but such generality won't be required in the following. Instead, we will require A to be the powerset $\mathcal{P}(X)$ of some set X. As usual in possible worlds models, X may be considered to be a set of worlds, and a proposition may be understood as the sets of worlds in which it is true. In this way, the models to be defined here adapt the model-theoretic approach of (standard) possible worlds models as defined in section 2.4.

As a result of these choices, the models to be defined can be thought of as specifying two sets of worlds, which need not be the same: X serves to give rise to the space of propositions, and D serves as the domain of quantification of world quantifiers. In the following, these will be distinguished by calling them "X-worlds" and "D-worlds," respectively. Models in which X-worlds coincide with D-worlds are especially natural, but various independence results can be established using models in which the two come apart. (When interpreted over models in which X-worlds coincide with D-worlds, $\mathcal{L}_w$ effectively becomes a version of hybrid logic; for more on hybrid logic, see Braüner (2017).)

Models also have to specify an interpretation of the connective $\triangleright$. This can be done using a two-place function T which maps any $d \in D$ and proposition $P \subseteq X$ to the proposition $T(d, P) \subseteq X$ that P is true in d.

Definition 7.9. *A* model *is a structure* $\mathfrak{M} = \langle X, D, T \rangle$ *such that X and D are non-empty sets and* $T : D \times \mathcal{P}(X) \to \mathcal{P}(X)$.

An assignment function *is a function a which maps every propositional variable p to a set $a(p) \subseteq X$ and every world variable w to an element $a(w)$ of D. As usual, let $a[P/p]$ be the assignment function which maps p to P and every $q \neq p$ to $a(q)$, and analogously for world variables.*

For each assignment function a, let $[\![\cdot]\!]_a$ be the function from $\mathcal{L}_w$ to $\mathcal{P}(X)$ satisfying the following conditions:

$$[\![p]\!]_a = a(p)$$

$$[\![\neg\varphi]\!]_a = X \backslash [\![\varphi]\!]_a$$

$$[\![\varphi \wedge \psi]\!]_a = [\![\varphi]\!]_a \cap [\![\psi]\!]_a$$

$$[\![\Box\varphi]\!]_a = \begin{cases} W & \textit{if } [\![\varphi]\!]_a = W \\ \varnothing & \textit{otherwise} \end{cases}$$

$$\llbracket \forall p \varphi \rrbracket_a = \bigcap_{P \subseteq X} \llbracket \varphi \rrbracket_{a[P/p]}$$

$$\llbracket \forall w \varphi \rrbracket_a = \bigcap_{d \in D} \llbracket \varphi \rrbracket_{a[d/w]}$$

$$\llbracket w = v \rrbracket_a = \begin{cases} W & \text{if } a(w) = a(v) \\ \varnothing & \text{otherwise} \end{cases}$$

$$\llbracket w \rhd \varphi \rrbracket_a = T(a(w), \llbracket \varphi \rrbracket_a)$$

A formula φ is true *relative to a model* $\mathfrak{M}$, $x \in X$, and assignment function a, written $\mathfrak{M}, x, a \vDash \varphi$, if $x \in \llbracket \varphi \rrbracket_a$.

φ is valid on $\mathfrak{M}$, written $\mathfrak{M} \vDash \varphi$, if $\mathfrak{M}, x, a \vDash \varphi$ for every $x \in X$ and assignment function a.

φ is valid, written $\vDash \varphi$, if $\mathfrak{M} \vDash \varphi$ for every model $\mathfrak{M}$.

Proposition 7.10 (Soundness). *If* $\vdash_w \varphi$ *then* $\vDash \varphi$.

Proof. Along the lines of the arguments in section 2.4, it is routine to show that every axiom is valid on every model $\mathfrak{M}$, and every rule preserves validity on $\mathfrak{M}$. An induction on the length of proofs shows that $\vdash_w \varphi$ only if $\mathfrak{M} \vDash \varphi$. □

Proposition 7.11. *None of* W1, W2 *and* W3 *is entailed by* WN, WP *in* $\vdash_w$.

Proof. By Proposition 7.10, it suffices to construct a model $\mathfrak{M}_n$ validating WN and WP, but not Wn (for $n \in \{1, 2, 3\}$). In all models to be specified, it is routine to verify that WN and WP are valid.

For W1, consider the model $\mathfrak{M}_1 = \langle X, D, T \rangle$ such that $X = \{0, 1\}$, $D = \mathcal{P}(X) \backslash \varnothing$, and $x \in T(d, P)$ iff $d \subseteq P$. To show that W1 is not valid on this model, let w be the D-world X. Then $\mathcal{A}w$ is false at each X-world x, If p is assigned $\{x\}$, then at x, p is true, but $w \rhd p$ is false.

For W2, consider the model $\mathfrak{M}_2 = \langle X, D, T \rangle$ specified as follows. Let $X = D = \{0, 1, 2\}$. For any $x \in X$, let π_x be $((x + 1 \bmod 3)(x + 2 \bmod 3))$, i.e., the permutation of $X = D$ mapping x to itself and the other two elements to each other. Let T be such that:

$$x \in T(d, P) \text{ iff } \begin{cases} d \in P & \text{if } x = d \\ \pi_x(d) \in P & \text{otherwise} \end{cases}$$

To show that W2 is not valid on this model, let w be the D-world 0 and p be the proposition $\{0\}$. Then $w \rhd p$ is true at 0 but false at 1.

For W3, consider the model $\mathfrak{M}_3 = \langle X, D, T \rangle$ such that $X = \{0\}$, $D = \{0, 1\}$ and $T(d, P) = P$. To show that W3 is not valid on this model, let w and v be 0 and 1, respectively. Then for any proposition p, $w \rhd p$ is true at 0 just in case $v \rhd p$ is, while $w = v$ is false.   □

Proposition 7.12. W *is independent in* $\vdash_w$, *in the sense that for every* $\varphi \in W$, $W\backslash\{\varphi\} \nvdash_w \varphi$.

Proof. To show that W0 cannot be derived from W1–3, consider the model $\mathfrak{M}_0 = \langle X, D, T \rangle$ such that $X = \{0, 1\}$, $D = \{0\}$ and $x \in T(d, P)$ iff $d \in P$. It is routine to show that this model validates W1–3. To show that it does not validate W0, let p be the proposition $\{1\}$ and w be the D-world 0; the claim follows from the fact that then, $w \rhd p$ is false (in both X-worlds).

The remaining claim, that each of W1–3 cannot be derived from the other principles of W, can be established using the models used in the proof of Proposition 7.11. All that needs to be shown is that the remaining principles are valid on the relevant model, which is routine.   □

It is worth noting that since the models used in this proof *validate* the relevant principles, the independence claim also holds for the necessitations of the members of W.

7.7 Supplement: Deductions

This appendix proves Theorem 7.6. Throughout, this section makes use of Lemma 7.3. A first lemma shows that according to W, worlds correspond one-to-one to maximal propositions, via $\mathcal{A}$:

Lemma 7.13. *The following are derivable from* W *in* $\vdash_w$:

(i) $\mathcal{A}w = \mathcal{A}v \rightarrow w = v$

(ii) $\mathrm{M}p \rightarrow \exists w(\mathcal{A}w = p)$

(iii) $\mathrm{M}\mathcal{A}w$

Proof. By three deductions indicated below. In the proof of (i), ND^w stands for the necessity of distinctness for worlds, which can be shown by a standard argument using the principles of S5; see Hughes and Cresswell (1996, p. 314).

(i):

(1)	$\Diamond(\mathcal{A}w \wedge \mathcal{A}w)$	W1
(2)	$\mathcal{A}w = \mathcal{A}v \rightarrow \Diamond(\mathcal{A}w \wedge \mathcal{A}v)$	1
(3)	$\mathcal{A}w \wedge \mathcal{A}v \rightarrow \forall p(w \rhd p \leftrightarrow v \rhd p)$	UI, UG
(4)	$\mathcal{A}w \wedge \mathcal{A}v \rightarrow w = v$	3, W3
(5)	$\mathcal{A}w = \mathcal{A}v \rightarrow \Diamond w = v$	2, 4
(6)	$\mathcal{A}w = \mathcal{A}v \rightarrow w = v$	5, ND^w

(ii):

(1)	$w \rhd p \rightarrow \Box w \rhd p$	W2
(2)	$\mathcal{A}w \dashv 3 \, (p \leftrightarrow w \rhd p)$	UI
(3)	$w \rhd p \rightarrow (\mathcal{A}w \dashv 3 \, p)$	1, 2
(4)	$\Diamond(\mathcal{A}w \wedge (p \leftrightarrow w \rhd p))$	W1, UI
(5)	$w \rhd p \rightarrow \Diamond(\mathcal{A}w \wedge p)$	1, 4
(6)	$w \rhd p \rightarrow \neg(p \dashv 3 \, \neg \mathcal{A}w)$	5
(7)	$\mathrm{M}p \wedge w \rhd p \rightarrow (p \dashv 3 \, \mathcal{A}w)$	6
(8)	$\mathrm{M}p \wedge w \rhd p \rightarrow \mathcal{A}w = p$	3, 7
(9)	$\mathrm{M}p \rightarrow \exists w(w \rhd p)$	Prop. 7.5, W0
(10)	$\mathrm{M}p \rightarrow \exists w(\mathcal{A}w = p)$	8, 9

(iii):

(1)	$\mathcal{A}w \dashv 3 \, (\neg p \leftrightarrow w \rhd \neg p)$	UI
(2)	$\neg(\mathcal{A}w \dashv 3 \, p) \rightarrow \Diamond(\mathcal{A}w \wedge \neg p)$	S5
(3)	$\Diamond(\mathcal{A}w \wedge \neg p) \rightarrow \Diamond w \rhd \neg p$	1
(4)	$\Diamond w \rhd \neg p \rightarrow \Box w \rhd \neg p$	W2
(5)	$\Box w \rhd \neg p \rightarrow (\mathcal{A}w \dashv 3 \, \neg p)$	1
(6)	$\neg(\mathcal{A}w \dashv 3 \, p) \rightarrow (\mathcal{A}w \dashv 3 \, \neg p)$	2–5
(7)	$\Diamond \mathcal{A}w \wedge ((\mathcal{A}w \dashv 3 \, p) \vee (\mathcal{A}w \dashv 3 \, \neg p))$	W1, 6
(8)	$\mathrm{M}\mathcal{A}w$	7, Prop. 7.5

$\square$

To account for new variables which occur freely, an auxiliary mapping will be needed, which replaces any such variable p_w by $\mathcal{A}w$. So let $\cdot^{|}$ be the mapping from $\mathcal{L}_p^+$ (the extension of $\mathcal{L}_p$ by the new propositional variables p_w for any world variable w) to $\mathcal{L}_w$ defined as follows:

$$\varphi^{|} := \varphi[\mathcal{A}w/p_w]\ldots[\mathcal{A}z/p_z]$$

where $w, \ldots, z$ are the free world variables in φ. With this, the following lemma can be established, from which Theorem 7.6 follows immediately:

Lemma 7.14. *For every $\mathcal{L}^w$-formula φ:*

$$W \vdash_w \varphi^{\S|} \leftrightarrow \varphi$$

Proof. By induction on the complexity of φ. Note that if by IH (induction hypothesis), $\varphi^{\S|} \leftrightarrow \varphi$ is provable, then so is $\varphi^{\S|} = \varphi$, using N. We consider the three non-trivial cases:

($\forall$) $(\forall p_w(Mp_w \to \varphi^\S))^{|} \leftrightarrow \forall w\varphi$ needs to be proven. This is established by the following two deductions, one for each direction:

(1)	$\forall p_w(Mp_w \to \varphi^\S) \to (M\mathcal{A}w \to \varphi^\S[\mathcal{A}w/p_w])$	UI
(2)	$\forall p_w(Mp_w \to \varphi^\S) \to \varphi^\S[\mathcal{A}w/p_w]$	1, Lemma 7.13(*iii*)
(3)	$(\forall p_w(Mp_w \to \varphi^\S))^{|} \to \varphi^{\S|}$	2, UG, UI
(4)	$(\forall p_w(Mp_w \to \varphi^\S))^{|} \to \forall w\varphi$	3, IH, UG^w

(1)	$\forall w\varphi \to \varphi^{\S|}$	UI^w, IH
(2)	$\forall w\varphi \to \forall w(\varphi^{\S|})$	1, UG^w
(3)	$\forall w(\varphi^\S[\mathcal{A}w/p_w]) \to ((\mathcal{A}w = p_w) \to \varphi^\S)$	UI^w, LL
(4)	$\forall w(\varphi^\S[\mathcal{A}w/p_w]) \wedge \neg\varphi^\S \to (\mathcal{A}w \neq p_w)$	3
(5)	$\forall w(\varphi^\S[\mathcal{A}w/p_w]) \wedge \neg\varphi^\S \to \forall w(\mathcal{A}w \neq p_w)$	4, UG^w
(6)	$\forall w(\varphi^\S[\mathcal{A}w/p_w]) \to (\exists w(\mathcal{A}w = p_w) \to \varphi^\S)$	5
(7)	$Mp_w \to \exists w(\mathcal{A}w = p_w)$	Lemma 7.13(*ii*)
(8)	$\forall w(\varphi^\S[\mathcal{A}w/p_w]) \to (Mp_w \to \varphi^\S)$	6, 7
(9)	$\forall w(\varphi^\S[\mathcal{A}w/p_w]) \to \forall p_w(Mp_w \to \varphi^\S)$	8, UG
(10)	$\forall w(\varphi^{\S|}) \to (\forall p_w(Mp_w \to \varphi^\S))^{|}$	9, UG, UI
(11)	$\forall w\varphi \to (\forall p_w(Mp_w \to \varphi^\S))^{|}$	2, 10

($\triangleright$) $(\mathcal{A}w \dashv3 \varphi^{\S})^{|} \leftrightarrow w \triangleright \varphi$ needs to be proven. By IH and UI, it suffices to prove $(\mathcal{A}w \dashv3 p) \leftrightarrow w \triangleright p$, which is established by the following deduction:

(1) $\mathcal{A}w \dashv3 (p \leftrightarrow w \triangleright p)$ UI
(2) $(\mathcal{A}w \dashv3 p) \to \Diamond(\mathcal{A}w \wedge p)$ W1
(3) $(\mathcal{A}w \dashv3 p) \to \Diamond w \triangleright p$ 1, 2
(4) $(\mathcal{A}w \dashv3 p) \to w \triangleright p$ 3, W2
(5) $w \triangleright p \to \Box w \triangleright p$ W2
(6) $w \triangleright p \to (\mathcal{A}w \dashv3 p)$ 1, 5
(7) $(\mathcal{A}w \dashv3 p) \leftrightarrow w \triangleright p$ 4, 6

($=$) $\mathcal{A}w = \mathcal{A}v \leftrightarrow w = v$ needs to be proven. This is immediate by Lemma 7.13(i).   □

8

Plural Propositional Quantification

8.1 Necessity without Worlds

The previous chapter concluded that whether talk of worlds can be taken
at face value depends on the truth of the principle ATOMICITY. This
principle states that necessarily, there is a true maximal proposition.
Since maximality is defined in modal terms, which have in turn been
defined in terms of identity, we can ask whether ATOMICITY is derivable
in $\vdash^=$ from the theory C5 motivated in previous chapters. It turns out
that the answer is negative.

Proposition 8.1. C5 $\not\vdash^=$ ATOMICITY.

The proof of this result is given in a supplementary section 8.6. It is
a straightforward extension of analogous proofs of the underivability of
ATOMICITY in $S_\Pi 5$, which was noted by Bull (1969); Kaplan (1970); and
Fine (1970). The fundamental idea is to interpret the relevant language
over an atomless complete Boolean algebra, where the elements of the
algebra serve as propositions. Such models were already mentioned in
the proof of Propositions 6.7. It can be shown that in such models, a
proposition is maximal iff it is atomic. Thus, if the algebra is atomless, no
proposition is atomic, which invalidates ATOMICITY.

The underivability of ATOMICITY shows that its truth does not follow
from claims motivated so far. But this doesn't mean that ATOMICITY is
not true. Recall that the 5 axiom was shown not to be derivable from
CLASSICISM in $\vdash^=$. Chapter 6 argued that an actuality operator can be
introduced, with which 5 can be derived. This chapter provides a similar
argument for the truth of ATOMICITY, by showing that ATOMICITY can
be derived once plural quantifiers over propositions are introduced. With
the results of the previous chapter, it follows that maximal propositions

The Foundations of Modality: From Propositions to Possible Worlds. Peter Fritz, Oxford University Press.
© Peter Fritz 2023. DOI: 10.1093/oso/9780192870025.003.0009

play the role of possible worlds. Via plural propositional quantifiers, it can thus be shown that there are possible worlds, and that they may be taken to be maximal propositions.

8.2 Adding Plural Quantifiers

The need for plural quantification arises in many contexts. One natural such context is the very argument which we will use below to establish ATOMICITY; for other examples, see Fritz et al. (2021); Hall (2021); Fritz (2022, 2023); and Kment (2022). Informally, the argument goes as follows. Consider first the truths (the propositions which are true), and then the proposition that they are all true. Every truth is true, so the claim that they (the truths) are all true is true as well. A universal claim strictly implies every one of its instances, so the claim that they (the truths) are all true strictly implies all of the truths, and so is maximal. Thus there is a true maximal proposition. This line of argument does not depend on any contingent assumptions, so the conclusion should hold necessarily. So, necessarily, there is a true maximal proposition, which is just what ATOMICITY claims.

This argument may seem suspicious, and the feeling of suspicion is likely going to increase if one considers the proposition that every truth is true, which is expressed by the following formula:

$$\forall p(p \rightarrow p)$$

This formula doesn't express any maximal proposition, but the tautologous one. However, $\forall p(p \rightarrow p)$ does not express the proposition appealed to in the argument just given: the argument considers the truths, i.e., the propositions p such that p, and the proposition that every one of *them* is true, rather than the proposition that every truth is true. This difference is crucial. To regiment the intended argument properly, the formal language needs to be extended by a way of quantifying *plurally* over propositions.

To illustrate the idea of plural quantification using a simpler example, consider the following claim:

There are some propositions which could each be true, but could not all be true together.

Such propositions are easy to come by. For example, for any contingent proposition p, p and $\neg p$ are individually possible but jointly impossible. In claims like the one just stated, the phrase "some propositions" clearly does not express singular quantification over propositions: it is non-sensical to assert of one proposition that it could be true individually but not collectively. "Some propositions" also plausibly does not express quantification over sets of propositions, as it is coherent to say that there are some propositions which do not form a set. (Indeed, in the present higher-order framework, it is natural to think that talk of sets and their members must be regimented using first-order quantifiers. With talk of propositions cashed out in terms of propositional quantifiers, talk about sets of propositions becomes incoherent.) In a literature going back to Boolos (1984), many have therefore considered phrases like "some propositions" to express a distinct form of *plural* quantification.

In $\mathcal{L}$, quantification over propositions is regimented using proposi-tional quantifiers rather than first-order quantifiers. Therefore, "some propositions" needs to be regimented using a plural version of propo-sitional quantification, rather than the more common plural version of first-order quantification. Technically, this poses no difficulties; in particular, standard axiomatizations of plural logic can be adapted straightforwardly.

Plural propositional quantifiers could be added to any of the languages considered here. For present purposes, it will suffice to add them to the propositionally quantified modal language $\mathcal{L}_p$. So, we add to $\mathcal{L}_p$ plural propositional variables $pp, qq, \ldots$, which can be bound by a universal quantifier $\forall$. Further, for any formula φ and plural propositional variable pp, we admit $\varphi \prec pp$ as a formula, expressing the claim that φ is one of pp. We extend $S_\Pi 5$ correspondingly. Apart from standard quantificational principles for plural propositional quantifiers, two axioms are required. First, a rigidity principle for "one of," according to which any proposition among some propositions is necessarily among them. In the context of S5, this means that the corresponding principle for "not one of," that any proposition not among some propositions is necessarily not among them is derivable as well. These plural rigidity principles are highly plausible, and although Hewitt (2012) questions them, they are widely endorsed in the literature; see Williamson (2003, p. 547, 2010, p. 699, 2013, pp. 245–249); Rumfitt (2005); Uzquiano (2011); Linnebo (2016); and Florio and Linnebo (2021, ch. 10). The second principle is a schematic comprehension principle, according to which for any condition φ, there are some propositions which are all and only the ones satisfying φ. Here, existential quantifiers are again defined as the dual of universal quantifiers.

Definition 8.2. *Let $\mathcal{L}_{pp}$ be the extension of $\mathcal{L}_p$ which adds to the recursive clauses 1–3 of $\mathcal{L}_p$ the following clause, using plural propositional variables $pp, qq \ldots$:*

(4) *If φ is a formula and pp is a plural propositional variable, then $\forall pp\varphi$ and $\varphi \prec pp$ are formulas.*

Let $\vdash_{pp}$ be the proof system extending $\vdash_{S\Pi5}$ by the following axiom schemas and rule:

(UI^p) $\forall pp\varphi \to \varphi[qq/pp]$
(UG^p) $\varphi \to \psi/\varphi \to \forall pp\psi$ *(pp not free in φ)*
(R$\prec$) $p \prec pp \to \Box(p \prec pp)$
(PC) $\exists pp\forall p(p \prec pp \leftrightarrow \varphi)$ *(pp not free in φ)*

The intended interpretation of plural propositional quantifiers is settled along the lines of the discussion in Chapter 1. First, the axiomatic principles of $\vdash_{pp}$ serve as stipulative logical principles which constrain the interpretation of these new quantifiers. Second, the interpretation of these quantifiers is further constrained by requiring them to correspond roughly to certain natural language phrases. In the case of plural propositional quantifiers, these phrases may include "for any zero or more propositions" and "for some zero or more propositions." Third, the interpretation of plural propositional quantifiers can be narrowed down by requiring them to relate to propositional quantifiers as standard plural quantifiers stand to first-order quantifiers.

It is worth remarking on a natural variant of plural propositional quantifiers which we could also have introduced. Instead of tying plural propositional quantifiers to phrases involving the qualifier "zero or more" (which is inspired by the discussion of Burgess and Rosen (1997, p. 155)), we could have tied them to simpler phrases omitting such a qualifier, such as "for any propositions" and "for some propositions." In English, such phrases plausibly carry an existential commitment, in the sense that for any propositions, there is some proposition among them. This can be motivated by noting that, e.g., from the claim that there *are* some propositions which are true, one naturally infers that there *is* some proposition which is true. On this variant introduction, the stipulative logical principles should therefore be amended as well: PC should be restricted to conditions which are satisfied, and a principle should be added stating that among any propositions, there is some proposition. PC should thus be replaced by the following two axiomatic principles:

(PC′) $\exists p\varphi \rightarrow \exists pp \forall p(p \prec pp \leftrightarrow \varphi)$ (*pp* not free in φ)
(PE) $\exists p(p \prec pp)$

Plausibly, both of these variants of the metasemantic constraints on $\mathcal{L}_{pp}$ can be used to determine a unique intended interpretation. In the following, the first option will be assumed. However, nothing of substance depends on this choice, as the central deductive argument to be given shortly can be carried out using the weaker comprehension principle PC′ instead of PC.

8.3 Worlds from Plural Quantification

The argument sketched above can now be vindicated. Recall that a crucial element of the argument is an appeal to the truths. In $\mathcal{L}_{pp}$, talk of the truths can be formalized by considering the propositions p such that p. The following instance of PC′ guarantees that there are those propositions, i.e., the propositions pp among which are just the truths, as there is some p such that p (e.g., $\top$):

$$\exists p\, p \rightarrow \exists pp \forall p(p \prec pp \leftrightarrow p)$$

The proposition that all pp are true can be stated as $\forall p(p \prec pp \rightarrow p)$. This already dispels some appearance of triviality, as this formula obviously differs from $\forall p(p \rightarrow p)$. In order to show that $\forall p(p \prec pp \rightarrow p)$ is a true maximal proposition when pp are the truths, it will be useful to introduce two defined notions: let T predicate of pluralities of propositions that they are all true, and $\mathbb{T}$ that they are all and only the truths:

$$Tpp := \forall p(p \prec pp \to p)$$
$$\mathbb{T}pp := \forall p(p \prec pp \leftrightarrow p)$$

With these, the next proposition shows that the informal argument can be turned into a rigorous deduction of ATOMICITY in $\vdash_{pp}$.

As noted above, ATOMICITY is not a theorem of $S_\Pi 5$. The next proposition therefore shows that $\vdash_{pp}$ does not extend $\vdash_{S_\Pi 5}$ conservatively. This is surprising, since $\vdash_{pp}$ appears to add to $\vdash_{S_\Pi 5}$ only axiomatic principles which govern the behavior of plural propositional quantifiers. This might raise the worry that $\vdash_{pp}$ is too strong, maybe even being inconsistent. In order to dispel this worry, the next proposition also shows that nothing stronger than ATOMICITY becomes derivable in $\vdash_{pp}$, in the sense that a formula φ of $\mathcal{L}_p$ is derivable in $\vdash_{pp}$ just in case it is derivable from ATOMICITY in $\vdash_{S_\Pi 5}$.

Proposition 8.3. *For all $\varphi \in \mathcal{L}_p$, $\vdash_{pp} \varphi$ iff* ATOMICITY $\vdash_{S_\Pi 5} \varphi$.

Proof. We first show that $\vdash_{pp} \varphi$ if ATOMICITY $\vdash_{S_\Pi 5} \varphi$. With a straightforward analog of Lemma 7.3 for $\vdash_{pp}$, it suffices to show that $\vdash_{pp}$ ATOMICITY. This is established by the following deduction:

(1)	$p \prec pp \to (Tpp \to p)$	UI
(2)	$p \prec pp \to (Tpp \dashv3 p)$	1, R$\prec$
(3)	$\mathbb{T}pp \to ((p \prec pp \leftrightarrow p) \wedge (\neg p \prec pp \leftrightarrow \neg p))$	UI
(4)	$\mathbb{T}pp \to \forall p(p \prec pp \vee \neg p \prec pp)$	3, UG
(5)	$\mathbb{T}pp \to \forall p(Tpp \dashv3 p \vee Tpp \dashv3 \neg p)$	2, 4
(6)	$\mathbb{T}pp \to \Diamond Tpp$	S5
(7)	$\mathbb{T}pp \to MTpp$	5, 6, Prop. 7.5
(8)	$\exists pp \mathbb{T}pp$	PC$'$
(9)	$\exists pp(MTpp \wedge Tpp)$	7, 8
(10)	$\forall p \neg(Mp \wedge p) \to \forall pp \neg(MTpp \wedge Tpp)$	UI, UG
(11)	$\exists p(Mp \wedge p)$	9, 10
(11)	ATOMICITY	11, N

The converse direction, showing that $\vdash_{pp} \varphi$ only if ATOMICITY $\vdash_{S_\Pi 5} \varphi$, is established by a model construction. Since this is a straightforward extension of well-known results, the details will only be sketched. Assume

ATOMICITY $\nvdash_{S_{\Pi 5}} \varphi$. Then by the completeness result of Kaplan (1970) and Fine (1970), there is a possible worlds model which falsifies φ. Such a model is given by a set of worlds, with necessity interpreted as truth in all worlds and propositional quantifiers ranging over sets of worlds. Extend the interpretation of formulas relative to such models to $\mathcal{L}_{pp}$, by letting plural propositional quantifiers range over sets of sets of worlds, and by letting $p \prec pp$ be true iff $a(p) \in a(pp)$, where a is the assignment function interpreting free variables. A routine induction on the length of proofs shows that $\vdash_{pp}$ is sound with respect to this interpretation, and so $\nvdash_{pp} \varphi$. □

The most important part of this proof is the deduction of ATOMICITY in $\vdash_{pp}$, which provides the central argument for the truth of ATOMICITY of this chapter. Variants of this argument have been given by a number of other authors. In particular, Gallin (1975, §11, see esp. theorem 11.5) shows in a higher-order modal logic that ATOMICITY follows from an axiom of comprehension for rigid properties. Fine (2005 [1977], pp. 136–137) informally presents a version of the argument using sets, which is worked out more formally in Fine (1980, pp. 192–195, see esp. theorem 37). A similar argument in a different formal setting can be found in Menzel and Zalta (2014). Gallin and Fine both argue model-theoretically, in the case of Gallin using a completeness result for a class of general ("Henkin") models; Menzel and Zalta reason deductively. The novelty of the present formulation is the appeal to plural propositional quantification. Another noteworthy variant of the plural argument can be formulated using the plural propositional abstracts of Fritz et al. (2021). Similar to λ-terms, these are plural propositional terms of the form $\pi p.\varphi$, standing for the propositions p such that $\varphi(p)$. The truths can then be expressed using $\pi p.p$; with this, the argument takes the form of showing that, necessarily, $\forall p(p \prec \pi p.p \to p)$ is a maximal truth.

8.4 World-Propositions and World-Stories

The deduction in the proof of Proposition 8.3 vindicates the argument sketched at the beginning of this chapter. Necessarily, considering the truths, the proposition that they are true is a maximal proposition. ATOMICITY follows. In the previous chapter, we saw that the truth of

ATOMICITY is the crucial assumption required for taking possible worlds talk at face value. Plural propositional quantifiers therefore vindicate appeals to possible wolds, and more specifically show that talk of possible worlds may be understood as talk of maximal propositions.

The upshot is that those who subscribe to the logical principles advocated here can defend their appeals to possible worlds in their theorizing by explicating worlds as maximal propositions. One might be tempted to assert more boldly that possible worlds *are* maximal propositions. But maximal propositions satisfying the role of possible worlds does not entail nothing else satisfying that role. And in fact, it is easy to show in $\vdash_{pp}$ that this role is also satisfied by certain pluralities of propositions. Call a plurality of propositions *pp maximal* if it is possible that *pp* are all and only the truths. Then maximal propositions correspond one-to-one to maximal pluralities of propositions, by the function which maps every maximal proposition to the plurality of propositions it strictly implies. Along these lines, it is routine to show that W is also satisfied by maximal *pluralities* of propositions. Since talk of possible worlds is highly theoretical, the question whether possible worlds are maximal propositions or maximal pluralities of propositions seems uninteresting, as long as both kinds of entities satisfy the required theoretical role.

The option of thinking of possible worlds as maximal pluralities of propositions makes salient a noteworthy aspects of the results established here. One might have thought that there is a coherent view which denies the existence of maximal propositions but countenances maximal *pluralities* of propositions. Along the model-theoretic ideas sketched in the argument for Proposition 8.1, one might have suggested to model this view by letting propositions be represented by elements of a complete but atomless Boolean algebra, and pluralities of propositions be represented by sets of these elements. But the results developed here show that such a model does not validate the intended claims: if there are maximal pluralities of propositions satisfying W, then the corresponding propositions—stating that every one of the relevant plurality is true—likewise satisfy W. In section 8.6, we consider such models in more detail, and note they do not validate all instances of the plural comprehension principle PC'. These observations show that even given a view such as

PROPOSITIONAL BOOLEANISM, which one might be tempted to sum up as requiring propositions to form a Boolean algebra, it makes a crucial difference whether we theorize about propositions in terms of a higher-order object language or in terms of standard mathematical reasoning about Boolean algebras. (For more on this point, see section 10.4 and Bacon and Dorr (forthcoming, §2).)

Maybe surprisingly, this section has shown that it is immaterial whether one thinks of worlds as propositions or as pluralities of propositions: there are propositions playing the role of worlds if and only if there are pluralities of propositions playing the role of worlds. And as argued here using plural propositional quantifiers, there are indeed both propositions and pluralities of propositions playing this role.

8.5 The Logic of Necessity, Once More

Apart from vindicating appeals to possible worlds, ATOMICITY allows us to determine further the logic of necessity. In section 6.5 we established the principles of $S_\Pi 5$, which is a conservative extension of S5. We can now add ATOMICITY to $S_\Pi 5$. Further, recall that section 4.1 argued for there being infinitely many propositions, which can be expressed using the $\mathcal{L}^=$-schema ∞ whose instances are of the form E_n, stating schematically that for each $n \in \mathbb{N}$, there are at least n propositions. Since all quantifiers involved in this schema are propositional, we can trans-late these instances into the propositionally quantified modal language $\mathcal{L}_p$ using the mapping $\cdot^\dagger$. Relying on an analog of Lemma 7.3, these claims can be added as axioms to our strengthened logic of necessity, containing only formulas which are true under every interpretation of their free propositional variables. We therefore define two strengthenings of $S_\Pi 5$:

Definition 8.4. *Let $S_\Pi 5At$ be the set of theorems of $\vdash_{S_\Pi 5}$ + ATOMICITY, the proof system obtained by adding ATOMICITY as an axiom to $\vdash_{S_\Pi 5}$. Let $S_\Pi 5At\infty$ be the set of theorems of $\vdash_{S_\Pi 5}$ + ATOMICITY $+\infty^\dagger$, the proof system obtained by adding ATOMICITY and the instances of $E_n^\dagger$ for $n \in \mathbb{N}$ as axioms to $\vdash_{S_\Pi 5}$.*

These logics are very well-behaved, and well-understood. Starting with $S_\Pi 5At$, Kaplan (1970) and Fine (1970) showed that this logic is sound and complete with respect to a natural model theory, namely the relational frames with a universal accessibility relation. Fine (1970, p. 342), Holliday (2019, p. 326), and Ding (2018) have made a number of further observations concerning $S_\Pi 5At\infty$. For present purposes, the most important ones can be summed up as follows.

First, working in propositional modal logic (without propositional quantifiers), Scroggs (1951) showed effectively that all proper extensions of S5 impose a finite limit on the number of propositions which can be distinguished using $\square$. According to the present view of propositional individuation, this means imposing a finite limit on the number of propositions. Correspondingly, it can be shown that $S_\Pi 5At\infty$ can alternatively be axiomatized by adding to $\vdash_{S_\Pi 5}$ +ATOMICITY the instances of the following axiom schema, stating that every non-theorem of S5 has a false instance:

$$\neg \bar{\forall} \varphi \quad (\text{where } \varphi \in \mathcal{L}_m \backslash S5)$$

This axiom schema was already proposed by Kripke (1959, p. 12), on which Bull (1969, p. 257) remarked, calling it a "perverse postulate."

Next, $S_\Pi 5At\infty$ is sound and complete with respect to any non-empty class of infinite relational frames with a universal accessibility relation. $S_\Pi 5At\infty$ is moreover negation-complete, meaning that for every closed formula of $\mathcal{L}_p$, $S_\Pi 5At\infty$ contains either it or its negation. So, the claim that $S_\Pi 5At\infty$ contains only formulas which are true under every interpretation of their free propositional variables can be strengthened to the claim that $S_\Pi 5At\infty$ contains all and only the formulas which are true under every interpretation of their free propositional variables. In the terminology of Williamson (2013), this means that the logic of metaphysical universality in the language $\mathcal{L}_p$ is precisely $S_\Pi 5At\infty$. $S_\Pi 5At\infty$ is also a conservative extension of S5, in the sense that the quantifier-free theorems of $S_\Pi 5At\infty$ are precisely the theorems of S5. Consequently, the quantifier-free formulas which are true under all interpretations of the propositional variables are precisely the theorems of S5; in Williamson's terminology, the logic of metaphysical universality in the language $\mathcal{L}_m$ is precisely S5.

8.6 Supplement: Independence

This supplementary section sketches a proof of C5 $\nvDash$ ATOMICITY. The proof uses models based on atomless complete Boolean algebras. For an introduction to the required algebraic notions, see Davey and Priestley (2002) and Givant and Halmos (2009). The argument is a minor variation of results which can be found in Gallin (1975); a number of details are therefore omitted.

In brief, a (complete) Boolean algebra can be understood as a special kind of order. A *partial order* is a pair $\langle S, \leq \rangle$ such that S is a set and $\leq$ a binary relation on S which is reflexive, transitive, and antisymmetric on S. A *lattice* is a partial order $\langle S, \leq \rangle$ such that any two elements of S have a least upper bound and a greatest lower bound in $\leq$. We write $s \sqcap t$ for the greatest lower bound of s and t, and $s \sqcup t$ for the least upper bound of s and t. A lattice is *complete* if every set $T \subseteq S$ has a least upper bound (or equivalently, greatest lower bound). We write $\sqcap T$ for the greatest lower bound of T, and $\sqcup T$ for the least upper bound of T. A lattice is *distributive* if for all $r, s, t \in S$:

$$r \sqcap (s \sqcup t) = (r \sqcap s) \sqcup (r \sqcap t)$$

A lattice is *bounded* if there are elements 0 and 1 such that for all $s \in S$, $0 \leq s \leq 1$. A lattice is *complemented* if there are elements 0 and 1 witnessing that it is bounded, such that for every $s \in S$, there is some $t \in S$ (the *complement of s*) for which $s \sqcap t = 0$ and $s \sqcup t = 1$. If a distributive lattice is complemented, then the complement of any $s \in S$ is unique, and we write $-s$ for it. A *(complete) Boolean algebra* is a (complete) complemented distributive lattice.

Possible worlds models can be thought of as based on special complete Boolean algebras. For any set W, the powerset $\mathcal{P}(W)$ together with the subset order $\subseteq$ on $\mathcal{P}(W)$ forms a complete Boolean algebra. In this case, the order-theoretic operations end up being the corresponding set-theoretic relations. For example, greatest lower bounds are intersections, and complements in the order-theoretic sense are relative complements in the set-theoretic sense. The distinguishing feature of such powerset algebras is the fact that they are *atomic*, which means that for every non-zero element s, there is some atom $a \leq s$, where an *atom* is a non-zero element a such that the only non-zero element $t \leq a$ is a itself. The atoms of a powerset algebra are the singleton sets, and can be thought of as the maximal propositions which play the role of worlds; this is the origin of the label ATOMICITY. The independence of ATOMICITY will therefore be shown using Boolean algebras which fail to be atomic, more specifically, using Boolean algebras which are *atomless*, which means that they do not contain any atoms. The algebras will be assumed to be complete; this will allow a straightforward interpretation of quantifiers.

We therefore start by showing how $\mathcal{L}^=$ can be interpreted on complete Boolean algebras. The elements of the algebra serve as propositions, and functions from n-tuples of propositions to propositions serve as n-ary modalities (similar to the construction in Definition 2.4). All that needs to be added is a valuation function which interprets the constants; with this, the recursive specification of the proposition expressed by a formula in a model is straightforward, using the operators provided by the Boolean algebra to interpret the logical connectives. The only further constant on which $\mathcal{L}^=$ imposes non-trivial constraints is the identity connective $=$; these constraints can be satisfied using a simple interpretation of $=$ which interprets any identity statement either as 0 or as 1:

Definition 8.5. *A Boolean model is a structure* $\mathfrak{M} = \langle B, \leq, V \rangle$, *satisfying the following requirements: First,* $\langle B, \leq \rangle$ *is complete Boolean algebra. For every* $n > 0$, *let:*

$$D_0 := B$$
$$D_n := B^{B^n}$$

Second, V *is a function mapping every constant of type n to an element of* D_n, *such that for all* $b_1, b_2 \in B$:

$$V(=)(\langle b_1, b_2 \rangle) = \begin{cases} 1 & \text{if } b_1 = b_2 \\ 0 & \text{otherwise} \end{cases}$$

An assignment function *is a function mapping every variable of type n to an element of* D_n. *For such an assignment function a, let* $[\![\cdot]\!]_a$ *be the function mapping every expression of* $\mathcal{L}^=$ *of type n to a member of* D_n *which satisfies the following conditions (omitting some Boolean connectives* *and the existential quantifier, which are treated analogously to the clauses provided here):*

$$[\![x]\!]_a := a(x)$$
$$[\![c]\!]_a := V(c)$$
$$[\![\neg\varphi]\!]_a := -[\![\varphi]\!]_a$$
$$[\![\varphi \wedge \psi]\!]_a := [\![\varphi]\!]_a \sqcap [\![\psi]\!]_a$$
$$[\![\forall x\varphi]\!]_a := \bigsqcap\{[\![\varphi]\!]_{a[o/x]} : o \in D_n\}$$
$$[\![\mu\varphi_1 \ldots \varphi_n]\!]_a := [\![\mu]\!]_a(\langle[\![\varphi_1]\!]_a, \ldots, [\![\varphi_n]\!]_a\rangle)$$
$$[\![\lambda\bar{p}.\varphi]\!]_a := \bar{d} \in D_0^n \mapsto [\![\varphi]\!]_{a[\bar{d}/\bar{p}]}$$

φ is valid on $\mathfrak{M}$ if $[\![\varphi]\!]_a = 1$ for every assignment function a.

φ is b-valid if φ is valid on every Boolean model.

Proposition 8.6. *If* C5 $\vdash^=$ φ *then φ is b-valid.*

Proof. Along the lines of Lemmas 2.10 and 2.12, we show that TAUT, λC, RI, and LL are *b*-valid, and *b*-validity is closed under MP. Similarly, we can show that the quantificational principles of $\vdash^=$ are *b*-valid (or preserve *b*-validity).

To show the validity of CLASSICISM, it suffices to show that *b*-validity is closed under RE. This follows from the fact that in any Boolean algebra, $b_1 \equiv b_2 = 1$ iff $b_1 = b_2$, where $\equiv$ is the Boolean operation corresponding to the biconditional. The argument is analogous to the deductive argument in Lemma 5.1. It remains to establish the *b*-validity of 5: Recall that 5 is $\vdash^=$-equivalent to $(p \neq \bot) \rightarrow ((p \neq \bot) = \top)$, so it suffices to establish the *b*-validity of the latter. If *p* is assigned 0, then $p \neq \bot$ is interpreted as 0. Otherwise, $p \neq \bot$ is interpreted as 1, whence $(p \neq \bot) = \top$ is interpreted as 1. In either case, 5 is interpreted as 1, as required.   □

Proof of Proposition 8.1. With the previous proposition, it suffices to show that ATOMICITY is not valid on Boolean models. This can be done by showing that ATOMICITY is interpreted as 0 in any Boolean model based on an atomless complete Boolean algebra. Since the underlying algebra is atomless, this follows from the claim that Mp is interpreted as 0 whenever *p* is not interpreted as an atom of the algebra. This claim in turn follows from the fact that $p \dashv_3 q$ is interpreted as 1 or 0, depending on whether the interpretation of p entails (in the sense of the algebraic order) the interpretation of q. □

As mentioned in section 8.4, although it is straightforward to interpret plural propositional quantifiers over Boolean models, such models won't all validate plural comprehension. First, in a given Boolean model, we can interpret plural propositional quantifiers as ranging over subsets of the domain of propositions B. Correspondingly, we can extend assignment functions to map plural propositional variables to subsets of B. With this, the interpretation clauses for plural propositional quantifiers are straightforward adaptations of those of other quantifiers, in particular:

$$\llbracket \forall pp\varphi \rrbracket_a = \bigcap \{\llbracket \varphi \rrbracket_{a[S/pp]} : S \subseteq B\}$$

In order to validate R≺, and in analogy with the clause for =, we can interpret ≺-formulas as either 1 or 0:

$$\llbracket \varphi \prec pp \rrbracket_a = \begin{cases} 1 & \text{if } \llbracket \varphi \rrbracket_a \in a(pp) \\ 0 & \text{otherwise} \end{cases}$$

We can now show that even though plural propositional quantifiers are interpreted as ranging over arbitrary sets of propositions in the model theory, some instances of plural comprehension fail to be b-valid. This is due to the fact that quantifiers are not interpreted in any straightforward truth-conditional way, but using greatest lower bounds and least upper bounds, which in this case leads to results which may be surprising.

In more detail, consider any model based on an atomless complete Boolean algebra $\langle B, \leq \rangle$. We show that the following comprehension instance is interpreted as 0 in this model:

$$\exists pp \forall p(p \prec pp \leftrightarrow p)$$

To do so, it suffices to show that $\forall p(p \prec pp \leftrightarrow p)$ is interpreted as 0 on any given assignment function a. Assume otherwise for contradiction. Then $\llbracket \forall p(p \prec pp \leftrightarrow p) \rrbracket_a = b$ for some $b > 0$. Since the algebra is atomless, there is some c such that $0 < c < b$. By the interpretation of universal quantifiers, $[\![p \prec pp \leftrightarrow p]\!]_{a[c/p]} \geq b$. We distinguish two cases. If $c \in a(pp)$, then $[\![p \prec pp]\!]_{a[c/p]} = 1$, whence $c \geq b$, contradicting $c < b$. If $c \notin a(pp)$, then $[\![p \prec pp]\!]_{a[c/p]} = 0$, whence $c \sqcap b = 0$, contradicting $0 < c < b$. Thus $[\![\forall p(p \prec pp \leftrightarrow p)]\!]_a = 0$ as required.

8.7 Supplement: Informal Completeness

This supplementary section makes good on a promise made in the previous chapter, which is to show that the theory of worlds W is informally complete. First, the question needs to be addressed as to what informal completeness amounts to: what does it mean that nothing is missing from W? A tempting answer is to say that W should capture the complete theoretical role of possible worlds. But theoretical roles are typically not explicitly articulated; instead, they emerge implicitly from a theoretical practice. And the theoretical practice in philosophy employing worlds is extensive and diverse. The prospects of articulating a single role of worlds which coheres with all the uses to which worlds have been put are slim. However, this does not preclude the possibility of assessing the informal completeness of W among a more limited domain of claims. Given the formal language $\mathcal{L}_w$ in which W is articulated, the most obvious such question is whether there are any sentences of $\mathcal{L}_w$ which should be added to W. It will now be argued that the answer to this question is negative: W is informally complete at least as far as the sentences of $\mathcal{L}_w$ go.

W of course leaves open whether a propositional variable p is true, but this is uninteresting. The interesting question is about the extent to which there are any *sentences* (formulas without free variables) which are not settled by W. If there are no such sentences, then W is *negation-complete*, which is to say that for every sentence φ, W entails either φ or $\neg\varphi$. If W is not negation-complete, then the questions not settled by it can be characterized by describing the negation-complete extensions of W.

With the results established above and additional results from the literature on propositional quantification in modal logic, we can provide such a characterization. By Theorem 7.6, quantification over worlds can be eliminated in favor of quantification over maximal propositions. Thus the negation-complete extensions of W in $\vdash_w$ correspond to the negation-complete extensions of $W^\$$ in $S_\Pi 5$. As shown above, $W^\$$ is equivalent to ATOMICITY in $S_\Pi 5$. And the consistent negation-complete extensions of ATOMICITY in $S_\Pi 5$ are known to be obtained by adding sentences determining the number of (maximal) propositions. This is an extension of the characterization of propositional modal logics extending S5 by Scroggs (1951), which was effectively observed by Fine (1970, p. 342); for a more detailed discussion, see Holliday (2019, p. 326).

The relevant sentences determining the number of maximal propositions can be formulated as stating the existence of at least n distinct maximal propositions, for any natural number n. One way of completing ATOMICITY is by stating that there are at least n maximal propositions, but not more, i.e., that there are not at least $n + 1$ maximal propositions, where $n > 0$. There is only one other way of completing ATOMICITY, which is by the claim that there are infinitely many propositions. This can be articulated by taking the set of sentences stating that there are at least n maximal propositions, for each natural number n. Due to the expressive limitations of $\mathcal{L}_w$, no distinctions can be made between cases in which the infinite cardinalities of maximal propositions differ. The required sets of sentences completing ATOMICITY can therefore be articulated as follows:

$$(\exists_n) \quad \exists p_1 \ldots \exists p_n \bigwedge_{i < j \leq n} (Mp_i \wedge Mp_j \wedge p_i \neq p_j)$$

$$C_n := \{\exists_n, \neg\exists_{n+1}\}$$
$$C_\infty := \{\exists_n : n \in \mathbb{N}\}$$

Proposition 8.7. *The negation-complete extensions of* W *consistent in* $\vdash_w$ *are axiomatized by* $W \cup C_n$, *for* $n > 0$, *and* $W \cup C_\infty$.

Proof. Let $\mathbb{N}^+_\infty = (\mathbb{N} \backslash \{0\}) \cup \{\infty\}$. We start by showing that for every $\mathcal{L}_w$-sentence φ and $\kappa \in \mathbb{N}^+_\infty$:

$$(*) \quad W, C_\kappa \vdash_w \varphi \text{ iff At}, C_\kappa \vdash_{S_\Pi 5} \varphi^\S$$

Assume first that At, $C_\kappa \vdash_{S_\Pi 5} \varphi^\S$. Then by Proposition 7.8, $W, C_\kappa \vdash_w \varphi^\S$, and so with Theorem 7.6, $W, C_\kappa \vdash_w \varphi$. Conversely, if At, $C_\kappa \nvdash_{S_\Pi 5} \varphi^\S$, then by the completeness theorem mentioned in the proof of Proposition 8.3, there is a possible worlds model, i.e., set W, which falsifies $\varphi^\S$. Then the

w-model $\langle X, D, T \rangle$ such that $X = D = W$ and $x \in T(d, P)$ iff $d \in P$ verifies the same $\mathcal{L}_p$-formulas as W. Thus this w-model falsifies $\varphi^\S$ but verifies the members of C_κ. By construction, it also verifies the members of W. With Theorem 7.6 again, this w-model falsifies φ, whence $W, C_\kappa \nvdash_w \varphi$.

Now consider $W \cup C_\kappa$, for any $\kappa \in \mathbb{N}^+_\infty$, with a view to proving that it is consistent and negation-complete in $\vdash_w$. By *, consistency follows using a possible worlds model, with the cardinality of worlds chosen according to κ. To establish negation-completeness, consider any $\varphi \in \mathcal{L}_w$. Since At $\cup$ C_κ is negation-complete in $S_\Pi 5$, as noted above, $\varphi^\S$ or $\neg\varphi^\S$ will be derivable from At $\cup$ C_κ in $\vdash_{S_\Pi 5}$. With *, it follows that φ or $\neg\varphi$ is derivable from $W \cup C_\kappa$ in $\vdash_w$, as required.

Finally, consider any consistent negation-complete set of sentences $\Gamma \subset \mathcal{L}_w$ extending W. Since the existence of a maximal proposition follows from W in $\vdash_w$, $\Gamma \vdash_w \exists_1$. If there is an $n \in \mathbb{N}$ such that $\Gamma \nvdash_w \exists_n$, then there is a first such n, and so by negation-completeness an $n \in \mathbb{N}$ such that both members of C_{n-1} follow from Γ in $\vdash_w$. Otherwise, every member of C_∞ follows from Γ in $\vdash_w$. Thus for some $\kappa \in \mathbb{N}^+_\infty$, every member of $W \cup C_\kappa$ follows from Γ in $\vdash_w$. As argued, $W \cup C_\kappa$ is consistent and negation-complete, whence Γ must be deductively equivalent to $W \cup C_\kappa$ in $\vdash_w$. $\qquad\qquad\square$

Proposition 8.7 shows that all that W leaves open in $\vdash_w$—as far as claims go which can be articulated in $\mathcal{L}_w$—is how many worlds there are. The existence of a particular number of worlds is plausibly not part of the general role of worlds. If the existence of, say, infinitely many worlds were to be included in the theoretical role of worlds, then W could straightforwardly be strengthened accordingly. In any case, once we strengthen $\vdash_w$ by including the axioms schema $\infty^\dagger$ requiring the number of propositions to be infinite—corresponding to the strengthening of $S_\Pi 5At$ to $S_\Pi 5At\infty$ in section 8.5—C_∞ will be derivable from W. With the assumption that there are infinitely many propositions, W can then be seen more straightforwardly to be informally complete by virtue of being negation-complete. In this stronger setting, W answers every question about worlds which can be articulated in $\mathcal{L}_w$.

What about questions which cannot be articulated in $\mathcal{L}_w$? The mode of argument used here to argue for informal completeness is limited to rather restrictive languages like W. In richer languages, the negation-complete extensions of many interesting axiomatic theories are often too numerous and diverse to admit of a neat characterization. A simple example can be given using the resources of the present chapter, as $\vdash_{pp}$ has extensions which are not recursively enumerable. This follows from the fact that when interpreted over possible worlds models in the most straightforward way, $\mathcal{L}_{pp}$ becomes inter-translatable with third-order monadic logic, which Tharp (1973) shows not to be recursively enumerable. In such richer settings, it appears that all one can do is proceed piecemeal, and consider particular principles and ask whether they follow from the theory of worlds developed so far.

PART V
CONCLUSION

9

A Coarse-Grained World-View

9.1 Summing Up

The previous chapter concludes the main argumentative trajectory toward a coarse-grained world-view in this book. Starting with the first chapter, this book has argued that debates about entities like propositions and modalities in metaphysics are usefully cast in terms of a higher-order language $\mathcal{L}$. Views on which propositions are individuated very finely were set aside due to a range of limitative results, building on the Russell-Myhill argument. CLASSICISM was identified as a natural starting point in an exploration of coarse-grained views, being distinguished in terms of simplicity and strength, without collapsing into EXTENSIONALISM. On this view, metaphysical necessity was argued to be the modality of being identical to the tautological proposition ⊤. This modality was shown to satisfy a natural quantified extension of the modal logic S4. Appealing to the intelligibility of an actuality operator, it was shown that this logic can be strengthened to a corresponding extension of S5. A theory of possible worlds was developed, and it was shown that it is satisfied if and only if the principle of ATOMICITY is true. This principle, stating that necessarily, there is a true maximal proposition, was justified by appealing to the intelligibility of plural quantification over propositions.

From the theory of possible worlds justified in this way, it follows that a proposition is necessary if and only if it is true in all possible worlds, and that propositions are identical just in case they are true in the same possible worlds. The resulting theory therefore vindicates a substantial amount of traditional intensional metaphysics. Although the picture arrived at in the end is very familiar, this book has taken a novel route toward it, explicitly identifying and motivating every one of its essential components.

The Foundations of Modality: From Propositions to Possible Worlds. Peter Fritz, Oxford University Press.

The resulting modal theory of propositions and their properties could be formulated in greater generality, and its formal presentation could be investigated in more detail. For example, it would be natural to consider a fuller higher-order language, such as the language of relational type theory $\mathcal{L}^*$ motivated in Chapter 1. However, this has effectively already been done by Gallin (1975, part II). Gallin considers a system which he calls $ML_P + C + EC$, defined in his book on p. 77. This system includes a principle of "extensional comprehension," which postulates the existence of a rigid property—a property which applies necessarily to everything it possibly applies to—for every defining condition. Higher-order quantification over such rigid properties can be used to simulate plural quantification, in the sense that an extension of Gallin's system by higher-order plural quantifiers proves that the proposition expressed by any formula involving plural quantifiers is identical to the proposition expressed by any corresponding formula in which plural quantifiers are replaced by the relevant higher-order quantifiers restricted to rigid properties. Many formal properties of higher-order extensions of the systems developed here can therefore be read off in a relatively simple way from Gallin's work.

The elimination of plural quantifiers using restricted higher-order quantifiers just sketched suggests the question whether the number of logical primitives can be reduced further, on the view endorsed here. In fact, it is easy to see that this is possible. For example, it is clear that the Boolean connectives of $\mathcal{L}$ could be reduced to any functionally complete set of connectives. Further discussion along these lines in closely related settings can be found in Muskens (2007), Benzmüller and Andrews (2019, §1.4), and Kneale and Kneale (1962, p. 522). However, for philosophical purposes, such reductions of primitives are often not very helpful. To illustrate this, consider again the case of plural quantification. It may be that plural quantifiers can in fact be eliminated using higher-order quantifiers restricted to rigid properties. But the possibility of this reduction depends on the truth of a principle along the lines of Gallin's principle of extensional comprehension, which entails that every property can be rigidified. And the best way of justifying this principle may be by appealing to the intelligibility of plural quantification.

9.2 Open Questions

The view developed here is strong, but it does not settle every question which can be asked in the language $\mathcal{L}$. Indeed, by the result of Tharp (1973) discussed in section 8.7, it is compatible with what has been argued for here that the truths in $\mathcal{L}$ are not recursively enumerable, so that such completeness may not be obtainable from an axiomatically presented theory. Given this potential incompleteness, it is natural to consider more restrictive fragments, such as the language $\mathcal{L}_p$ of propositionally quantified modal logic. As discussed in section 8.5, the logic $S_{\Pi}5At\infty$ argued for in this book settles every question which can be formulated in the language $\mathcal{L}_p$.

Many of the natural questions in richer languages like $\mathcal{L}$ left open here have a distinctly mathematical character. These include questions about the particular infinity of propositions, and about versions of the axiom of choice and the continuum hypothesis which can be formulated in higher-order logic; for examples of such formulations, see Shapiro (1991, pp. 105–107). While these questions are interesting for particular applications of higher-order logic, especially in the philosophy of mathematics, they are not obviously important for most metaphysical discussions.

In metaphysical contexts, it appears more important to consider the application of the theory to particular metaphysical questions. Since possible worlds theorizing has been highly influential in metaphysics over the last sixty years, this is territory which has been rather well-explored, albeit mainly in an informal way. In the end, a proper assessment of any metaphysical view has to be carried out comparatively, by comparing it to its most promising alternatives. Few alternative approaches to propositional individuation have been developed as thoroughly as the possible worlds approach. In a sense, then, the most promising way of making progress on assessing and potentially motivating the present coarse-grained viewpoint is to develop its competitors more thoroughly, in order to be able to carry out a proper comparison.

Developing such alternative viewpoints is not the topic of this book, but the next chapter outlines a range of options. The remaining supplementary sections to this chapter reply to two common and general objections to two central aspects of the view motivated here. The first concerns the individuation of propositions by the possible worlds in which they are true. The second concerns the use of higher-order languages.

9.3 Supplement: Challenges to Possible Worlds Accounts of Propositions

Possible worlds accounts of propositions—on which propositions are identical just in case they are true in the same possible worlds—have often been considered to be problematic, for a number of reasons. Many of these reasons concern propositional attitudes like knowledge and belief, and are based on the observation that a relatively coarse individuation of propositions conflicts with various intuitive judgements about attitudes. These issues include versions of Frege's puzzle, as well as the problem of logical omniscience. Since they were already discussed at length in Chapter 3, there is no need to go into them again.

However, there are also some other challenges to possible worlds accounts of propositions. A prominent one is due to Kaplan (1995). For a simple version of it, consider the attitude of entertaining a proposition, or, in Kaplan's phrase, *querying* the proposition. However coarsely or finely individuated propositions are, it seems at least *prima facie* plausible that every proposition could be uniquely queried (either by a specific agent, or by any agent—the difference doesn't matter much here). Kaplan formulates this principle as follows:

(A) $\forall p \Diamond \forall q (Qq \leftrightarrow p = q)$

This is a formula of $\mathcal{L}^{=Q}$, where Q is a unary operator informally read as standing for being queried. Another example for the interpretation of Q which makes Kaplan's principle A intuitively compelling, suggested by (Yli-Vakkuri, 2018), concerns the interpretation of a given sentence letter l: it seems at least *prima facie* plausible that every proposition could serve as the unique interpretation of the letter l.

Kaplan notes that A cannot be true in any standard possible worlds model. This is routine to verify using the model theory in section 2.4, using any standard model in which the truth of $\Diamond \varphi$ requires φ to be true in some world. Informally, the argument can be seen to be an application of Cantor's theorem. Since propositions are modeled as sets of possible worlds, Cantor's theorem tells us that there are more propositions than possible worlds. Therefore, there cannot be, for every proposition p, a world in which it is uniquely queried.

Since this argument appeals essentially to the individuation of propositions in terms of the possible worlds in which they are true, this argument initially seems to be a problem for the possible worlds account of propositions. However, the failure of A can be established deductively, from significantly weaker assumptions. This is an immediate consequence of a formalized version of the Epimenides paradox, which can be found, e.g., in Hilbert and Ackermann (1938) and Prior (1961). In the present setting, we can show that CLASSICISM entails that A is false on any interpretation of Q:

Theorem 9.1. CLASSICISM $\vdash^= \neg\exists m\forall p\Diamond\forall q(mq \leftrightarrow p = q)$

Proof. The following informal reasoning can straightforwardly be carried out in $\vdash^=$. Consider any m. We show that φ defined as $\forall r(mr \to \neg r)$ is a counterexample to the claim $\Diamond\forall q(mq \leftrightarrow \varphi = q)$.

Assume for contradiction that $\forall q(mq \leftrightarrow \varphi = q)$. Then $m\varphi$. If φ, then $\forall r(mr \to \neg r)$, whence $\neg\varphi$, which is a contradiction. If $\neg\varphi$, then there is some r such that mr and r. Since $\neg\varphi$ and r, $\varphi \neq r$. So mr and $\varphi \neq r$, which contradicts the assumption.

Having derived $\neg\forall q(mq \leftrightarrow \varphi = q)$, CLASSICISM allows us to necessitate, from which we obtain $\neg\Diamond\forall q(mq \leftrightarrow \varphi = q)$. By existential introduction $\exists p\neg\Diamond\forall q(mq \leftrightarrow p = q)$, from which the claim follows by universal generalization and the duality of quantifiers. $\square$

What Kaplan thought of as a problem for possible world semantics is therefore a much more general phenomenon. Given that it is established here as a theorem of CLASSICISM, one might take it to be a problem for CLASSICISM instead. But note that there was only one, implicit, appeal to CLASSICISM in the argument, in an application of the rule of necessitation. Whoever considers Kaplan's problem to be pressing must have some modality in mind for which A is plausible. As long as this notion counts the theorems of $\vdash^=$—all of which are true by stipulation—as necessary, a version of Theorem 9.1 will apply.

What to make of all of this is a difficult question, and not one it will be possible to settle here; for recent further discussion, see Bacon et al. (2016) and Ding and Holliday (2020). Two brief remarks will have to suffice. First, the logical limits on how finely propositions are individuated provide general reasons against regarding intuitions about attitude ascriptions to be very reliable. Second, recall from section 3.4 that the argument from attitude ascriptions to a structured individuation of propositions is much more plausible in the case of propositions expressed by closed formulas, rather than open formulas on arbitrary interpretations of the free variables. There, it was noted that with plural propositional quantifiers, a contradiction can be obtained just from CLOSED STRUCTURE, i.e., the closed instances of the schema STRUCTURE. Fritz et al. (2021, pp. 1281–1282) note that it is, in contrast, not clear that a contradiction can be obtained from a suitable schematic variant of A, restricted to closed instances. If a consistency result is possible along these lines, it would do much to relieve the tension posed by results like Theorem 9.1.

9.4 Supplement: Challenges to Higher-Order Languages

Chapter 1 noted that higher-order languages are sometimes criticized for not providing any formalization of a number of *prima facie* intelligible claims. An especially compelling form of this criticism notes that advocates of higher-order languages often seem to be engaging in precisely this kind of talk—talk which cannot straightforwardly be formalized in higher-order languages—when discussing higher-order languages. This is taken to show that there is something incoherent about higher-order languages. Arguments of this kind have been formulated by a number of authors including Menzel (1993, p. 66) and Bealer (1994, pp. 145–146), the latter in response to a lengthy and critical review of Bealer (1982) by Anderson (1987). Historically, the argument has roots in the famous problem of the concept horse discussed by Frege (1892a), as well as related discussion in the Tractatus of Wittgenstein (1921); for more on the connections between these two, see Proops (2013). An early statement of the criticism can also be found in Gödel (1984 [1944], p. 466).

To discuss objections along these lines, it will be best to focus on specific examples of statements which are taken to be problematic. We will focus on two such examples here. For the first example, recall that in line with the rejection of SELF-APPLICATION, $\mathcal{L}^*$ does not allow for a property term to be applied to itself. One might object that this is an implausible restriction. In particular, one might claim that the use of higher-order languages in regimenting talk of properties amounts to attempting to solve the paradox of the property of properties which don't apply to themselves merely by forbidding one to talk about it.

This is not a particularly compelling form of the criticism. The failure of SELF-APPLICATION in $\mathcal{L}^*$ is not due to any *restriction*. To see this clearly, consider $\mathcal{L}_1$, the language of first-order logic. In $\mathcal{L}_1$, a predicate F can be applied to an individual variable x to produce a formula Fx. The fact that one cannot construct formulas of the form FF or xx is not an implausible restriction of $\mathcal{L}_1$, but simply a consequence of unproblematic distinctions between its syntactic categories. $\mathcal{L}^*$ merely builds on these distinctions in its successive expansions of $\mathcal{L}_1$. In Chapter 1, each step in this process of expansion was motivated by desiderata of a language of properties, and a range of metasemantic factors were appealed to with the aim of endowing the additional syntactic constructions with meaning. There is therefore nothing unnaturally *restrictive* about $\mathcal{L}^*$.

To be sure, one may argue that $\mathcal{L}^*$ is *impoverished*, and that there are certain important claims that a useful language for talking about properties should be able to express but which $\mathcal{L}^*$ fails to express. But in this form, the argument against $\mathcal{L}^*$ seems to amount to nothing more than an affirmation of the desideratum of SELF-APPLICATION. And this desideratum was set aside here for very good reasons: it is inconsistent with some more important desiderata.

A second version of the objection to higher-order language starts from Frege's problem of the concept horse. As many advocates of higher-order logic have argued, the lesson to draw from this problem is that a semantic theory of a higher-order language has to be couched in a higher-order language itself. This was already noted by Russell (1956 [1918]); for recent discussion, see Williamson (2003), as well as Trueman (2015) and Jones (2016). According to this reply to the objection, the semantic theory of a higher-order language like $\mathcal{L}^*$ should regiment the notion of what a term of type t expresses in terms of an expression which takes as an argument an individual term a and a term b of type t, stating that the term denoted by a expresses what b expresses. Thus, the relevant semantic theory is not formulated using a single notion of expressing, but a type-indexed one, formulated using a separate expression for every type.

In developing such a semantic theory, the proponent of higher-order languages will of course not write down infinitely many sentences. It appears, then, that they require some form of generalization across types. However, the quantifiers of a language like $\mathcal{L}^*$ are restricted to particular types. This version of the objection as been put forward by Linnebo (2006), who illustrates it using the following example:

(UNIQUE EXISTENCE)   "Every expression of every syntactic category has a semantic value which is unique, not just within a particular type, but across all types."

Since this seems to be a correct principle concerning the semantics of higher-order languages, the proponent of such languages seems to face the challenge of stating it properly, which appears to require the kind of quantification across types, sometimes called "type-neutral quantification," which languages like $\mathcal{L}^*$ do not provide.

However, on reflection, it is not so easy to say what it takes to meet this challenge. To start, UNIQUE EXISTENCE is presumably a necessary truth. Thus, the proposition intended to be expressed is necessary, and so—at least on the coarse-grained theory of propositions motivated here—the unique necessary proposition. And this proposition is easily expressed using any tautology. Of course, this is a cheap reply. The advocate of the coarse-grained individuation of propositions has to concede that there is an important sense of expressing a claim which goes beyond expressing the right proposition. One could hardly claim to have answered every open mathematical question by stating that it is raining or not raining.

But once it becomes clear that the challenge is not to express the right proposition, but to convey it in the right way (presumably in a way which human interlocutors find illuminating), the challenge to express the relevant insight in higher-order language becomes less pressing. The formal languages introduced above were explicitly motivated by appeal to metaphysical theorizing about properties, alongside propositions and relations. There was no presumption for it to serve as a language for all purposes. In particular, it was no desideratum for this language to be able to formulate its own semantic theory. Furthermore, the so-called semantic paradoxes provide good independent reasons to think that theorizing about the semantics of one language must in general be done in a different language.

Setting aside the particular claim of UNIQUE EXISTENCE, one might respond to this reply by challenging the proponent of higher-order languages to formally regiment a semantic theory of their preferred higher-order language in some (other) language. And indeed, it might be problematic if there is no way of extending a given higher-order language X in a way which allows the formulation of the semantic theory of X. But no reason has been given for thinking that doing so is impossible. For example, note that the semantics of any n-th order language $\mathcal{L}_n$ might be given in $\mathcal{L}_m$, for m suitably larger than n. (Jones (2016, pp. 160–162) responds to Linnebo's challenge roughly along these lines.) The question does remain how one might state, in suitable generality, a semantic theory of the full language $\mathcal{L}^*$. But just as the semantics of $\mathcal{L}_n$ may be formulated in $\mathcal{L}_m$, the semantics of $\mathcal{L}^*$ may be stated in a suitably (transfinitarily) higher-order language. The details of this will likely be complex, and developing such a language is beyond the scope of this book. The next section discusses some options informally.

9.5 Supplement: Infinite Arity

Infinitary extensions of the type hierarchy of higher-order languages have been discussed in a number of places in the recent literature on higher-order logic, including Linnebo and Rayo (2012), Williamson (2013, pp. 235–240) and Krämer (2017); in the logical literature, there are already technical explorations in, e.g., Andrews (1965). The mentioned authors all use a so-called *cumulative* extension of the type hierarchy. On this approach, terms of infinitary types take as arguments terms of multiple different types. To illustrate this, one may extend $\mathcal{L}^*$ by a unary cumulative type, where expressions of this type take as arguments expressions of all types of $\mathcal{L}^*$.

Conceptually, this is a very novel kind of expansion. The new infinitary type cannot be introduced along the lines of the introduction of the finite types of $\mathcal{L}^*$ in Chapter 1: it cannot be introduced by successive introductions of variables for complex terms and λ-terms formed by binding these variables. (For related criticisms of cumulative type theory, see Button and Trueman (2022).) However, there is an alternative form of infinitary type theory, which was already suggested by Fine (1977, p. 144). Although it is technically less well explored, it is conceptually more conservative. This is the option to allow λ to bind not just finite sequences of variables, but infinite sequences as well. More generally, this approach admits relation symbols of infinite arities up to some specific ordinal limit.

Developing such infinite arity type theories is a task for another occasion. But there is a worry about appealing to relations of infinite arity which might already suggest itself at this informal level. One might claim that doing so leads to an infinite hierarchy of higher-order languages. To formulate a semantic theory of $\mathcal{L}^*$, one might move to a further language admitting relation symbols of arity up to ω. But to do the same for this further language, one would then have to move to yet another language, admitting relations of even higher arity, and so on. In this case, one might consider the proper object of semantic theorizing to be the *hierarchy* of languages, rather than a single language. To do semantics in suitable generality, it might be argued that one would have to talk about the semantic values of expressions of the languages in this whole hierarchy. And no single language has the resources to talk about all of these expressions.

But this suggestion should just be resisted. There is no reason to concede that in theorizing about the semantics of higher-order languages, one is ever concerned with any kind of completed hierarchy, rather than one particular higher-order language (often left implicit, as in the statement of UNIQUE EXISTENCE above). This point is illustrated well by considering how the syntax of the relevant higher-order languages may be formulated. Proponents of higher-order logic may well doubt that there is a unique intended interpretation of set-theoretic language. They might therefore prefer to formulate their formal theory of syntax of higher-order languages in these very higher-order languages, rather than set theory.

Since the resources of higher-order languages get richer the more types are admitted, the higher one gets in the hierarchy of languages the greater is the length of sequences one has at one's disposal for formulating a syntax for a higher-order language. In this sense, it is plausible that infinite arities provide a straightforward way of making sense of an open-endedness of the hierarchy of types. With this open-endedness, there is no sense to be made of reading any semantic claim as pertaining to a complete hierarchy of languages, rather than a particular language.

10
Looking Ahead

10.1 Backtracking

The last chapter concluded that to properly assess the orthodox modal metaphysics motivated here, its alternatives need to be developed more fully. Such alternatives can take many forms. This section discusses a few examples, without attempting to provide a comprehensive survey. The examples will be structured according to the main choice points discussed in this book, retracing the argumentative steps taken in previous chapters.

Rejecting Plural Quantification. The last step taken above was the introduction of plural propositional quantifiers, which led to an endorsement of the principle ATOMICITY. This principle was used to vindicate appeals to possible worlds. One might, however, reject the required instances of plural propositional comprehension, and so also reject the general existence of rigid properties and maximal propositions. (This could take the form either of rejecting plural propositional quantification outright or of rejecting the relevant instances of plural comprehension. The latter option might be motivated by a kind of indefinite extensibility view; see, e.g., Florio and Linnebo (2021, ch. 12).) As noted in section 8.6, models of such theories can be constructed using atomless complete Boolean algebras to serve as spaces of propositions. Interestingly, the model-theoretic results of Gallin (1975, §15) along these lines already show that such a rejection of full plural higher-order comprehension is compatible with retaining full plural individual comprehension, even assuming the existence of infinitely many individuals.

A more concrete example of such a view can be given based on *possibilities* or *states of affairs*. On this kind of view, the truth of propositions in possible worlds is replaced by the verification of propositions

The Foundations of Modality: From Propositions to Possible Worlds. Peter Fritz, Oxford University Press.
© Peter Fritz 2023. DOI: 10.1093/oso/9780192870025.003.0011

in states. This basic idea can be cashed out in many different forms; see, e.g., van Fraassen (1969); Humberstone (1981); Rumfitt (2015); and Fine (2017a,b). One particularly interesting model-theoretic approach takes states to be ordered by a relation of refinement and constructs propositions as sets of states satisfying certain closure properties formulated in terms of refinement. The resulting spaces of propositions can be shown to be structurally equivalent to complete Boolean algebras. Such a connection is in fact exploited in the use of Boolean-valued models for forcing in set theory, as presented in Bell (2005); this connection is developed in more detail in Holliday (forthcoming).

Rejecting "Actually."   One step earlier, one might object to the appeal to an actuality operator in motivating the necessity of distinctness and possibility. It was already noted in Chapter 6 that it is controversial whether there is any reading of "actually" in ordinary language with the required behavior. One might argue that there is in fact no property of propositions satisfying the stipulative logical principles A1 and A2, and so that the proposed introduction of the connective "@" fails. On such a view, one might hold that some instances of the principle 5 of the necessity of possibility are false, from which it follows that the necessity of distinctness fails as well (5 effectively being an instance of ND, given the definition of $\Box$ adopted here).

A more concrete example of such a view can be motivated by a Humean thought according to which there are no necessary connections between distinct fundamental entities. Taking non-logical constants to stand for fundamental entities, this can be cashed out by requiring $\Diamond\varphi$ to be true for every closed formula φ which is consistent in a suitable deductive system. Classicism in particular is amenable to such a viewpoint, as Classicism is consistent with the schematic claim that the truth of every closed formula consistent with Classicism is possible. In fact, this is equivalent to a schematic claim that every distinctness claim consistent with Classicism is true. For proofs of these claims and motivation of such views, see the discussion of "maximalist classicism" in Bacon (2020) and Bacon and Dorr (forthcoming).

Rejecting Classicism.   Classicism itself may of course also be rejected. Classicism was arrived at as the natural end-point of successively stronger schemas of equations corresponding to biconditionals provable in certain systems. Along these lines, CLASSICISM was meant to be motivated as a naturally strong and simple coarse-grained view formulated in $\mathcal{L}^=$. CLASSICISM was noted to lead to a number of controversial consequences, which may well be taken to be reasons to reject the view.

One example of such a consequence is PROPOSITIONAL NECESSITISM, the view that it is necessary what propositions there are. Many authors have found it plausible that it is contingent what individuals there are, and that some propositions are existentially dependent on such contingently existing individuals; the relevant propositions therefore exist only contingently themselves. Such a view of PROPOSITIONAL CONTINGENTISM is explored by Fritz (2016) and Fritz and Goodman (2016). Endorsing this view requires weakening CLASSICISM.

Rejecting the Operationalization of Simplicity and Strength.   The first step in motivating CLASSICISM was to show that among theories consisting of equations whose flanking terms are taken from the restrictive language $\mathcal{L}^B$ of propositional logic, the unique strongest consistent theory is PROPOSITIONAL BOOLEANISM, containing exactly those equations which correspond to tautologous biconditionals. This result was extended to the language $\mathcal{L}^{BQ}$ obtained by adding propositional quantifiers. However, it was noted that the result cannot be extended to other languages, such as the extension of $\mathcal{L}^B$ by the identity symbol. CLASSICISM was therefore motivated not by such a result identifying the relevant theory uniquely in terms of strength, but by extending the weaker equational theories in an analogous way. Those who reject CLASSICISM for the reasons mentioned in the previous paragraph may therefore still endorse PROPOSITIONAL BOOLEANISM (and its extension by propositional quantifiers), which was shown to be uniquely distinguished in terms of strength.

Going back one more step, one might also question these more restricted equational theories. In motivating them as being distinguished in terms of simplicity and strength, it is crucial that simplicity is operationalized in syntactic terms, as being formulated as a set of equations whose flanking terms are taken from a particular fragment of the language $\mathcal{L}$. It might be argued that this operationalization does not correctly capture the notion of simplicity which should guide theory choice.

In arguing against the relevant equational theories, one might point to particular natural distinctions which they rule out. One such distinction concerns the individuals a proposition is *about*. One might hold that the proposition j that Jupiter is a planet is about Jupiter, whereas the proposition m that Mercury is a planet is not about Jupiter, but about Mercury. So far, no conflict arises. But one may further hold that such aboutness is preserved under Boolean combinations. That is, one might hold that any Boolean combination of propositions is about just those individuals the combined propositions are about. Thus, $j \wedge \neg j$ is about Jupiter and not Mercury, whereas $m \wedge \neg m$ is about Mercury and not Jupiter. However, this is inconsistent with the equation $(p \wedge \neg p) = (q \wedge \neg q)$ which corresponds to the tautological biconditional $(p \wedge \neg p) \leftrightarrow (q \wedge \neg q)$. An example of a theory of propositional identity which accommodates such a notion of aboutness is mentioned by Goodman (2019); for more on recent work on aboutness, see Hawke (2018).

Rejecting Coarseness.   Beyond rejecting the particular coarse-grained theories of propositional identity developed above, one might reject more generally the idea that the lesson to be learned from the Russell-Myhill argument is that one should look for relatively coarse-grained theories of propositional individuation. Instead, one might propose to identify independent metaphysical considerations to provide a motivated standard of propositional identity which neither falls into the inconsistencies of naive theorizing about structured propositions, nor leads to any recognizably coarse-grained theory.

An example for such a view is the *no-circularity* view of Dorr (2016, §8), which can be motivated as making sense of the notion of real definition in metaphysics. Another example starts from the notion of metaphysical ground. As noted in section 3.3, naive principles of immediate grounding lead to an inconsistently fine-grained individuation of propositions. But one might weaken these principles, and try to find theories of propositional individuation which are coarse-grained enough to avoid inconsistency, while being fine-grained enough to accommodate a substantial number of non-trivial grounding principles. One possible route to such a view starts from a different version of the state-space approach to propositions mentioned above, which Fine (2012) has used to provide models for grounding.

Rejecting $\mathcal{L}$. The alternatives considered so far take the form of theories formulated in the language $\mathcal{L}$. More fundamentally, one might also reject the very language $\mathcal{L}$ used here, along with the fuller relationally typed language $\mathcal{L}^*$. This need not take the form of rejecting the argument for higher-order languages entirely. It might, for example, just take the form of arguing for a different kind of higher-order language.

For example, faced with the inconsistency of standard grounding principles in $\vdash^=$ discussed in section 3.3, a staunch proponent of grounding may not abandon the notion of grounding in metaphysics, but instead conclude that whatever uses $\mathcal{L}^=$ has, a different language should be used to capture talk of grounding. For instance, they might treat quantifiers not as variable binders but as higher-order predicates, and reject the principle of λC of λ-conversion. For some initial developments of such a view, see Fritz (2020) and Goodman (2023).

More radically, one might adopt a different type structure. A well-known alternative to relational type theory is functional type theory. On certain assumptions, there are plausible ways of translating between relational and functional type theory, as discussed by Dorr (2016). However, these assumptions are not uncontroversial. There are well-motivated views which deny the relevant assumptions, such as the theory of structured propositions of Bacon (2023). Bacon uses a functional type theory which imposes some substantial limitations on the formation rules for types. These restrictions can be used to block the derivation of the Russell-Myhill result, although the consistency of Bacon's theory is left as a conjecture.

Rejecting Higher-Order Languages. Finally, one may reject the argument for higher-order languages entirely. In particular, one might disagree with the assessment that the desideratum of SELF-APPLICATION should be given up. There is a wide variety of alternative options. For example, this could take the form of rejecting classical logic, as Field (2004) does. Or one might reject the transparency principle, as Bealer (1982) does.

10.2 Metaphysics and Metasemantics

The seven choice points just discussed divide naturally into two kinds: the first five concern the endorsement of different theories in the language $\mathcal{L}$, whereas the last two concern the adoption of different formal languages. In discussing views of the second kind, which take a substantially different approach to regimenting talk of propositions, properties, relations, and modalities, it is important to note that the various options need not be seen as competitors, except in the sense that for any given assertion, one must choose a language in which to make that assertion. It is perfectly coherent to suppose that there are several different meaningful formal languages of very different characters, all of which improve on the relevant ways of talking in natural languages. Indeed, the different formal languages may even be compatible in the sense that they can combined into a single formal language.

This is not to say that the dispute between different formal languages for reasoning about propositions, properties, relations, and modalities in metaphysics is bound to be verbal. For example, it is certainly conceivable that the introduction of $\mathcal{L}^*$ (and $\mathcal{L}$) is not successful. This may be because the various metasemantic stipulations made in Chapter 1 to settle the interpretation of the language are incompatible. Alternatively, it may be because these stipulations underdetermine the interpretation of the language to such an extent that it remains uninterpreted, or at least highly ambiguous, a worry which is urged by Clarke-Doane and McCarthy (2022). As they note, the worry is not just about underdetermination: even if a unique interpretation has been settled, it would be worrisome if this was chosen arbitrarily from a range of candidates. For example, if the interpretation of $\mathcal{L}$ is highly sensitive to minor changes in the metasemantic stipulations, the question of truth in $\mathcal{L}$ loses much of its interest.

It is now time to respond to these worries. First, it is important to note that these are extremely general worries, which apply to any new linguistic devices one might attempt to introduce. If they are genuine, they threaten any attempt to improve on natural languages using meaningful formal languages. The higher-order quantifiers employed here may seem to be especially worrying, but recall that any consistent formalization of property talk will have to come apart in some ways from the naive ordinary judgements which lead to inconsistency, and so will be subject to similar metasemantic worries. Indeed, in one way, the present approach has been extremely conservative in ideological respects: the theory of modality and possible worlds has been constructed on the basis of a language containing only Boolean connectives, quantifiers, and identity. Despite metasemantic concerns, we cannot do any theorizing without a language in which to do the theorizing. The approach chosen here of not admitting distinctively metaphysical notions such as metaphysical necessity and metaphysical grounding as ideological primitives already concedes substantial ground to metasemantic worries. Unless we resist those worries at some point, they will paralyze any theorizing. A language containing only the distinctively logical resources of Boolean connectives, quantifiers, and identity seems like a very natural point at which to draw the line and to resist these worries.

Second, one might acknowledge that metasemantic worries have to be resisted at some point, and that certain ideology simply has to be defended as being in good standing, but take the lesson to be that before any theorizing can occur, much more semantic and metasemantic ground clearing is requiring. This book has also conceded a lot of ground to this worry, but at some point—arguably reached here—it should be resisted as well: even if the good standing of languages like $\mathcal{L}$ has not been established beyond any reasonable doubt, it does not follow that more work should be done to establish the language as meaningful before using it in its intended applications in metaphysics and philosophy more generally. Trying to put these languages on an indubitable footing before using them in their intended applications is not always going to be the best approach.

To illustrate this, consider a more down-to-earth example from physics. Imagine that a new kind of elementary particle appears to be detected in some experiments. Scientists introduce a new term with the intention of referring to this new kind. Is the term in good standing? At this stage of inquiry, this is hard to tell. If a new kind of particle was detected, it may well be that the term succeeds in picking out this kind. If the appearance of a new kind of particle is due to some malfunctioning of instruments, then it may well be that the introduced term is meaningless. In various scenarios involving multiple kinds of particles, it may be that the introduced term is ambiguous, at least initially.

Being aware of these semantic possibilities, how should the physicists involved in this research proceed? The best way to make progress is clearly not to worry too much about the semantics of the new term, and rather to continue to investigate the new phenomena. In formulating theories of the supposedly newly discovered particle, it is best simply to use the new term and to assume that it is in good standing. If such theories are unsuccessful, and the phenomena receive satisfying alternative explanations which do not appeal to new kinds of particles, there is good reason to abandon the term. Alternatively, if one of the theories involving the new term ends up being successful, this provides good reason to think that the introduced term succeeds in picking out a new kind of elementary particle. Indeed, it may be that the resulting theory ends up disambiguating a term whose meaning was initially insufficiently constrained. The best way to resolve worries about the semantic status of new theoretical vocabulary is therefore often to set aside these worries, to investigate the subject matter using the new terminology, and to trust that semantic and metasemantic defects will resolve themselves in the course of inquiry.

Metaphysics isn't physics. But like physics, metaphysics develops theories of various phenomena which are evaluated and compared according to a range of criteria, and improved over time. It is plausible that the metasemantic processes which govern the theoretical terms of metaphysics are analogous to those in the case of physics. This suggests that the best way to make progress in theorizing about propositions, properties, relations, and modalities is simply to proceed with carrying out metaphysical investigations in formal languages such as the one discussed here. It is important to acknowledge that these formal languages may well be defective in various ways. But this doesn't mean that one shouldn't work with them. In fact, even if they turn out to be meaningless in the end, this realization may constitute genuine progress.

10.3 Hyperintensional Metaphysics

This book has motivated a familiar intensional approach to metaphysics. Nevertheless, the previous chapter argued that some of the most promising lines of investigation in metaphysics are concerned with the development of alternative viewpoints. Despite appearances to the contrary, the position advocated here is therefore in agreement with a recent trend in metaphysics sometimes called *hyperintensional metaphysics*; see Nolan (2014) and Berto and Nolan (2021). This label is applied to general metaphysical inquiries which explicitly reject, or at least question, the assumption that propositions (and properties and relations) are individuated in terms of being necessarily co-extensive. Furthermore, the present investigation can shed some light on this general project, and help to advance it.

Consider by way of example a prominent example of hyperintensional metaphysics, the case of grounding. Advocates of metaphysical grounding often hold that propositions which are metaphysically necessarily equivalent may nevertheless be distinguished in terms of the relation of grounding, and so must be distinct. As discussed in section 3.3, this is often motivated by logical grounding principles which contradict CLASSICISM. This case serves to illustrate two important points.

First, once CLASSICISM is rejected, it may no longer be guaranteed that there is a distinguished single tautological proposition ⊤. This is of course very much intended, as a proponent of grounding may want to distinguish propositions $p \vee \neg p$ and $q \vee \neg q$: the former may be considered to be grounded just in p, and the latter just in q. But with the absence of a distinguished tautological proposition, and the failure of CLASSICISM more generally, the argument given above that there is a singe broadest necessity breaks down. One therefore faces anew the question how metaphysical necessity should be identified, and the skeptical arguments discussed above which question the assumption that current metaphysical theorizing singles out one distinguished modality. It is of course possible to develop theories of broadest necessities without the assumption of CLASSICISM, and first steps in this area are done by Bacon and Zeng (2022). But doing so is not a straightforward matter. Similarly, even if CLASSICISM is assumed, but the resources of the actuality operator and plural propositional quantifiers are rejected, it becomes questionable whether sense can be made of talk of possible worlds.

Although the alternative conceptions of metaphysics with which proponents of hyperintensional metaphysics engage might be very interesting, the label *hyperintensional* is therefore potentially misleading. The fundamental conceptual resources in the relevant enterprise are that of identity and whatever else is involved in the particular project (such as, in the case just discussed, grounding). In some instances, the notion of metaphysical necessity may simply fall by the wayside. Consequently, the very notion of hyperintensionality—of drawing a distinction between intensionally equivalent entities—may therefore come to be rejected as well. The relevant approaches to metaphysics by proponents of hyperintensional metaphysics may therefore be hyperintensional only in the sense of being incompatible with INTENSIONALISM, and not in the sense of endorsing the existence of cases which could adequately be described as exhibiting the phenomenon of hyperintensionality. The label *postmodal metaphysics* employed by Sider (2020, ch. 1) may therefore be more fitting. What is central to these inquiries is arguably not so much hyperintensionality, in the sense of drawing distinctions between what is intensionally equivalent, but simply the notion of identity, and various fine-grained views regarding it. This motivates adopting the same kind of identity-first approach to metaphysics which this book has followed, and which is advocated by Dorr (2016).

Second, the case of grounding shows the dangers and difficulties of endorsing fine-grained distinctions in metaphysics. Such fine distinctions are liable to lead to inconsistency via the Russell-Myhill argument and its many variations. This reinforces the importance of focusing on identity. It is not enough to provide some specific cases in which it seems as if metaphysical distinctions are drawn which violate CLASSICISM. General theories need to be developed to show that the relevant underlying ideas lead to theories of individuation which are logically consistent.

10.4 Models and Theories

The inconsistency of various fine-grained theories of propositional individuation shows that an important part of developing a viable metaphysical view is simply showing that it is consistent. Doing so is often harder and more subtle than is appreciated. This is because to show that a certain theory of propositions is consistent by the usual model-theoretic means, one must find models for the *theory of* propositions, and so *talk of* propositions, rather than just the propositions themselves.

It is worth illustrating this general point with a more concrete example. An important case is the standard model-theoretic treatments of structured propositions, as found in, e.g., Lewis (1970) and Cresswell (1985). Models of structured propositions in this tradition often start by assigning coarse-grained meanings to all expressions of a formal language along standard lines, for example using possible worlds models. Fine-grained meanings are then identified by assigning to each syntactically complex expression a tree with the same structure as the expression, with the coarse-grained meaning of each atomic constituent attached to the corresponding leaf of the tree.

Although such constructions are well-defined in the set theory in which the models are developed, they do not establish the consistency of views according to which proposition are structured. The problem is that these models do not interpret any language which provides quantifiers over the fine-grained meanings themselves, let alone quantifiers over pluralities or properties of such meanings. These models therefore do not show that *theories* of structured propositions are consistent, and this is what is at issue when considering the viability of the view that propositions are structured. It is therefore easy to underestimate the difficulty of developing alternatives to INTENSIONALISM. The relevant model-theoretic ideas may still provide interesting consistency arguments. For example, it can be shown along these lines how it is possible to find, on a coarse-grained metaphysics, certain kinds of proxies for structured propositions; see (Fritz 2021). But there are substantial limits on the role these proxies can play.

The question which theories of propositional identity are consistent cannot neatly be separated from the choice of the language in which these theories are formulated. This is illustrated by the theory developed by Bacon (2023), which suggests that restricting the type formation rules of higher-order logic may ensure the consistency of a structured individuation of propositions. To a certain extent, formal languages and the theories of propositional identity formulated in them are therefore best seen as package views. This does not mean that in articulating a position, one must adopt at the outset a complete view about what counts as a legitimate ideological resource; it suffices to start with some language one assumes to be legitimate, and in which one can start formulating a view of propositional identity. Consider again the use of higher-order languages in the present book. First, a general higher-order language $\mathcal{L}^*$ was introduced, but it was noted that a fragment $\mathcal{L}$ suffices for many purposes in developing a theory of propositions and their properties. Restricting the discussion to $\mathcal{L}$ simplified much of the presentation. Second, resources outside of $\mathcal{L}^*$ played an important role at various points, such as the use of plural propositional quantifiers in vindicating the notion of a possible world. There are therefore good reasons to remain flexible in one's choice of language.

Bibliography

Robert Merrihew Adams. Theories of actuality. *Noûs*, 8:211–231, 1974.

Robert Merrihew Adams. Actualism and thisness. *Synthese*, 49:3–41, 1981.

C. Anthony Anderson. Bealer's *Quality and Concept*: *Journal of Philosophical Logic*, 16:115–164, 1987.

Peter B. Andrews. A reduction of the axioms for the theory of propositional types. *Fundamenta Mathematicae*, 52:345–350, 1963.

Peter B. Andrews. *A Transfinite Type Theory with Type Variables*. Amsterdam: North-Holland, 1965.

Andrew Bacon. The broadest necessity. *Journal of Philosophical Logic*, 47:733–783, 2018a.

Andrew Bacon. *Vagueness and Thought*. Oxford: Oxford University Press, 2018b.

Andrew Bacon. Logical combinatorialism. *Philosophical Review*, 129:537–589, 2020.

Andrew Bacon. A theory of structured propositions. *Philosophical Review*, 132: 173–238, 2023.

Andrew Bacon and Cian Dorr. Classicism. In Peter Fritz and Nicholas K. Jones, editors, *Higher-Order Metaphysics*. Oxford: Oxford University Press, forthcoming.

Andrew Bacon and Jin Zeng. A theory of necessities. *Journal of Philosophical Logic*, 51:151–199, 2022.

Andrew Bacon, John Hawthorne, and Gabriel Uzquiano. Higher-order free logic and the Prior-Kaplan paradox. *Canadian Journal of Philosophy*, 46:493–541, 2016.

Ruth C. Barcan. A functional calculus of first order based on strict implication. *The Journal of Symbolic Logic*, 11:1–16, 1946.

Jon Barwise and John Perry. Shifting situations and shaken attitudes: An interview with Barwise and Perry. *Linguistics and Philosophy*, 8:105–161, 1985.

George Bealer. *Quality and Concept*. Oxford: Clarendon Press, 1982.

George Bealer. Property theory: The type-free approach v. the Church approach. *Journal of Philosophical Logic*, 23:139–171, 1994.

John L. Bell. *Set Theory: Boolean-Valued Models and Independence Proofs*. Oxford: Clarendon Press, 2005.

Christoph Benzmüller and Peter Andrews. Church's Type Theory. In Edward N. Zalta, editor, *The Stanford Encyclopedia of Philosophy*. Metaphysics Research Lab, Stanford University, 2019.

Francesco Berto and Daniel Nolan. Hyperintensionality. In Edward N. Zalta, editor, *The Stanford Encyclopedia of Philosophy*. Metaphysics Research Lab, Stanford University, 2021.

George Boolos. To be is to be a value of a variable (or to be some values of some variables). *The Journal of Philosophy*, 81:430–449, 1984.

Torben Braüner. Hybrid logic. In Edward N. Zalta, editor, *The Stanford Encyclopedia of Philosophy*. Metaphysics Research Lab, Stanford University, 2017.

R. A. Bull. On modal logics with propositional quantifiers. *The Journal of Symbolic Logic*, 34:257–263, 1969.

John P. Burgess. Tarski's tort. In *Mathematics, Models, and Modality: Selected Philosophical Essays*, pages 149–168. Cambridge: Cambridge University Press, 2008.

John P. Burgess and Gideon Rosen. *A Subject with No Object*. Oxford: Clarendon Press, 1997.

Tim Button and Robert Trueman. Against cumulative type theory. *The Review of Symbolic Logic*, 15:907–949, 2022.

Herman Cappelen. *Fixing Language: An Essay on Conceptual Engineering*. Oxford: Oxford University Press, 2018.

Rudolf Carnap. *Abriß der Logistik*. Vienna: Springer, 1929.

David J. Chalmers, David Manley, and Ryan Wasserman, editors. *Metametaphysics: New Essays on the Foundations of Ontology*. Oxford: Oxford University Press, 2009.

Alonzo Church. A formulation of the simple theory of types. *The Journal of Symbolic Logic*, 5:56–68, 1940.

Alonzo Church. Carnap's introduction to semantics. *The Philosophical Review*, 52: 298–304, 1943.

Alonzo Church. A formulation of the logic of sense and denotation. In Paul Henle, Horace M. Kallen, and Suzanne K. Langer, editors, *Structure, Method, and Meaning: Essays in Honor of Henry M. Scheffer*, pages 3–24. New York: Liberal Arts Press, 1951.

Alonzo Church. *Introduction to Mathematical Logic*. Princeton, NJ: Princeton University Press, 1956.

Justin Clarke-Doane. Modal objectivity. *Noûs*, 53:266–295, 2019.

Justin Clarke-Doane. Metaphysical and absolute possibility. *Synthese*, 198 (Suppl 8): S1861–S1872, 2021.

Justin Clarke-Doane and William McCarthy. Modal pluralism and higher-order logic. *Philosophical Perspectives*, 36:31–58, 2022.

M. J. Cresswell. Another basis for S4. *Logique et Analyse*, 8:191–195, 1965.

M. J. Cresswell. Functions of propositions. *The Journal of Symbolic Logic*, 31:545–560, 1966.

M. J. Cresswell. Propositional identity. *Logique et Analyse*, 10:283–292, 1967.

M. J. Cresswell. Second-order intensional logic. *Zeitschrift für mathematische Logik und Grundlagen der Mathematik*, 18:297–320, 1972.

M. J. Cresswell. *Structured Meanings*. Cambridge, MA: MIT Press, 1985.

John N. Crossley and Lloyd Humberstone. The logic of "actually." *Reports on Mathematical Logic*, 8:11–29, 1977.

Charles B. Daniels and James B. Freeman. Classical second-order intensional logic with maximal propositions. *Journal of Philosophical Logic*, 6:1–31, 1977.

B. A. Davey and H. A. Priestley. *Introduction to Lattices and Order*. Cambridge: Cambridge University Press, second edition, 2002.

Yifeng Ding. On the logics with propositional quantifiers extending S5Π. In Guram Bezhanishvili, Giovanna D'Agostino, George Metcalfe, and Thomas Studer, editors, *Advances in Modal Logic*, volume 12, pages 219–235. London: College Publications, 2018.

Yifeng Ding and Wesley H. Holliday. Another problem in possible world semantics. In Nicola Olivetti, Rineke Verbrugge, and Sara Negri, editors, *Advances in Modal Logic*, volume 13. London: College Publications, 2020.

Cian Dorr. Transparency and the context-sensitivity of attitude reports. In Manuel García-Carpintero and Genoveva Martí, editors, *Empty Representations: Reference and Non-Existence*, pages 25–66. Oxford: Oxford University Press, 2014.

Cian Dorr. To be F is to be G. *Philosophical Perspectives*, 30:39–134, 2016.

Cian Dorr and Jeremy Goodman. Diamonds are forever. *Noûs*, 54:632–665, 2020.

Cian Dorr and John Hawthorne. Naturalness. In Karen Bennett and Dean W. Zimmerman, editors, *Oxford Studies in Metaphysics*, Volume 8, pages 3–77. Oxford: Oxford University Press, 2014.

Cian Dorr, John Hawthorne, and Juhani Yli-Vakkuri. *The Bounds of Possibility: Puzzles of Modal Variation*. Oxford: Oxford University Press, 2021.

Hartry Field. *Realism, Mathematics, and Modality*. Oxford: Basil Blackwell, 1989.

Hartry Field. The consistency of the naïve theory of properties. *The Philosophical Quarterly*, 54:78–104, 2004.

Kit Fine. Propositional quantifiers in modal logic. *Theoria*, 36:336–346, 1970.

Kit Fine. Properties, propositions and sets. *Journal of Philosophical Logic*, 6:135–191, 1977.

Kit Fine. First-order modal theories II: Propositions. *Studia Logica*, 39:159–202, 1980.

Kit Fine. The question of realism. *Philosophers' Imprint*, 1:1–30, 2001.

Kit Fine. *Modality and Tense: Philosophical Papers*. Oxford: Oxford University Press, 2005.

Kit Fine. Prior on the construction of possible worlds and instants. In *Modality and Tense: Philosophical Papers*, pages 133–175. Oxford: Oxford University Press, 2005 [1977]. First published as postscript to *Worlds, Times and Selves* (with A. N. Prior), pp. 116–168. London: Duckworth, 1977.

Kit Fine. Guide to ground. In Fabrice Correia and Benjamin Schnieder, editors, *Metaphysical Grounding*, pages 37–80. Cambridge: Cambridge University Press, 2012.

Kit Fine. A theory of truthmaker content I: Conjunction, disjunction and negation. *Journal of Philosophical Logic*, 46:625–674, 2017a.

Kit Fine. A theory of truthmaker content II: Subject-matter, common content, remainder and ground. *Journal of Philosophical Logic*, 46:675–702, 2017b.

Kit Fine. Truthmaker semantics. In Bob Hale, Crispin Wright, and Alexander Miller, editors, *A Companion to the Philosophy of Language*, pages 556–577. Chichester: Wiley, second edition, 2017c.

Salvatore Florio and Øystein Linnebo. *The Many and the One: A Philosophical Study of Plural Logic*. Oxford: Oxford University Press, 2021.

Peter Forrest. Ways worlds could be. *Australasian Journal of Philosophy*, 64:15–24, 1986.

Gottlob Frege. *Begriffsschrift, eine der arithmetischen nachgebildete Formelsprache des reinen Denkens*. Halle a. S.: Louis Nebert, 1879.

Gottlob Frege. Über Begriff und Gegenstand. *Vierteljahresschrift für wissenschaftliche Philosophie*, 16:192–205, 1892a.

Gottlob Frege. Über Sinn und Bedeutung. *Zeitschrift für Philosophie und philosophische Kritik*, NF 100:25–50, 1892b.

Peter Fritz. Propositional contingentism. *The Review of Symbolic Logic*, 9:123–142, 2016.

Peter Fritz. A purely recombinatorial puzzle. *Noûs*, 51:547–564, 2017.

Peter Fritz. On higher-order logical grounds. *Analysis*, 80:656–666, 2020.

Peter Fritz. Structure by proxy, with an application to grounding. *Synthese*, 198: 6045–6063, 2021.

Peter Fritz. Ground and grain. *Philosophy and Phenomenological Research*, 105: 299–330, 2022.

Peter Fritz. Operands and instances. *The Review of Symbolic Logic*, 16:188–209, 2023.

Peter Fritz. *Propositional Quantifiers*. Elements in Philosophy and Logic. Cambridge: Cambridge University Press, forthcoming.

Peter Fritz and Jeremy Goodman. Higher-order contingentism, part 1: Closure and generation. *Journal of Philosophical Logic*, 45:645–695, 2016.

Peter Fritz, Harvey Lederman, and Gabriel Uzquiano. Closed structure. *Journal of Philosophical Logic*, 50:1249–1291, 2021.

Daniel Gallin. *Intensional and Higher-Order Modal Logic*. Amsterdam: North-Holland, 1975.

Steven Givant and Paul Halmos. *Introduction to Boolean Algebras*. New York: Springer, 2009.

Jeremy Goodman. Agglomerative algebras. *Journal of Philosophical Logic*, 48:631–648, 2019.

Jeremy Goodman. Grounding generalizations. *Journal of Philosophical Logic*, 52:821–858, 2023.

Jeremy Goodman. Higher-order logic as metaphysics. In Peter Fritz and Nicholas K. Jones, editors, *Higher-Order Metaphysics*. Oxford: Oxford University Press, forthcoming.

Kurt Gödel. Über formal unentscheidbare Sätze der Principia Mathematica und verwandter Systeme I. *Monatshefte für Mathematik und Physik*, 38:173–198, 1931.

Kurt Gödel. Russell's mathematical logic. In Paul Benacerraf and Hilary Putnam, editors, *Philosophy of Mathematics: Selected Readings*, pages 447–469. Cambridge: Cambridge University Press, second edition, 1984 [1944]. First published in Paul A. Schilpp, editor, *The Philosophy of Bertrand Russell*, pages 125–153. Evanston, Ill: Northwestern University, 1944.

Andrzej Grzegorczyk. The systems of Leśniewski in relation to contemporary logical research. *Studia Logica*, 3:77–97, 1955.

Bob Hale. Absolute necessities. *Philosophical Perspectives*, 10:93–117, 1996.

Bob Hale. *Necessary Beings: An Essay on Ontology, Modality, and the Relations Between Them*. Oxford: Oxford University Press, 2013.

Geoffrey Hall. Indefinite extensibility and the principle of sufficient reason. *Philosophical Studies*, 178:471–492, 2021.

Fredrik Haraldsen. On what *Actually* is. *Erkenntnis*, 80:643–656, 2015.

Peter Hawke. Theories of aboutness. *Australasian Journal of Philosophy*, 96:697–723, 2018.

Leon Henkin. A theory of propositional types. *Fundamenta Mathematicae*, 52:323–344, 1963.

Simon Thomas Hewitt. Modalising plurals. *Journal of Philosophical Logic*, 41:853–875, 2012.

D. Hilbert and W. Ackermann. *Grundzüge der theoretischen Logik*. Berlin: Springer-Verlag, second edition, 1938.

Harold T. Hodes. Why ramify? *Notre Dame Journal of Formal Logic*, 56:379–415, 2015.

Wesley H. Holliday. A note on algebraic semantics for S5 with propositional quantifiers. *Notre Dame Journal of Formal Logic*, 60:311–332, 2019.

Wesley H. Holliday. Possibility semantics. In Melvin Fitting, editor, *Selected Topics from Contemporary Logics*, pages 363–476. London: College Publications, 2021.

Wesley H. Holliday. Possibility frames and forcing for modal logic. *The Australasian Journal of Logic*, forthcoming.

G. E. Hughes and M. J. Cresswell. *A New Introduction to Modal Logic*. London: Routledge, 1996.

Lloyd Humberstone. From worlds to possibilities. *Journal of Philosophical Logic*, 10:313–339, 1981.

Nicholas K. Jones. A higher-order solution to the problem of the concept *Horse*. *Ergo*, 3:132–166, 2016.

David Kaplan. S5 with quantifiable propositional variables. *The Journal of Symbolic Logic*, 35:355, 1970.

David Kaplan. On the logic of demonstratives. *Journal of Philosophical Logic*, 8:81–98, 1978.

David Kaplan. Demonstratives. In Joseph Almog, John Perry, and Howard Wettstein, editors, *Themes from Kaplan*, pages 481–563. Oxford: Oxford University Press, 1989 [1977]. Completed and circulated in mimeograph in the published form in 1977.

David Kaplan. A problem in possible-world semantics. In Walter Sinnott-Armstrong, Diana Raffman, and Nicholas Asher, editors, *Modality, Morality, and Belief*, pages 41–52. Cambridge: Cambridge University Press, 1995.

John T. Kearns. Modal semantics without possible worlds. *The Journal of Symbolic Logic*, 46:77–86, 1981.

Jeffrey C. King. Structured propositions. In Edward N. Zalta, editor, *The Stanford Encyclopedia of Philosophy*. Metaphysics Research Lab, Stanford University, 2019.

Boris Kment. Russell-Myhill and grounding. *Analysis*, 82:49–60, 2022.

William Kneale and Martha Kneale. *The Development of Logic*. Oxford: Clarendon Press, 1962.

Stephan Krämer. Everything, and then some. *Mind*, 126:499–528, 2017.

Saul A. Kripke. A completeness theorem in modal logic. *The Journal of Symbolic Logic*, 24:1–14, 1959.

Saul A. Kripke. Semantical analysis of modal logic I: Normal modal propositional calculi. *Zeitschrift für mathematische Logik und Grundlagen der Mathematik*, 9:67–96, 1963.

Saul A. Kripke. *Naming and Necessity*. Cambridge, MA: Harvard University Press, 1980 [1972]. First published in *Semantics of Natural Language*, edited by Donald Davidson and Gilbert Harman, pages 253–355, 763–769. Dordrecht: D. Reidel, 1972.

Harvey Lederman. Higher-order metaphysics and propositional attitudes. In Peter Fritz and Nicholas K. Jones, editors, *Higher-Order Metaphysics*. Oxford: Oxford University Press, forthcoming.

Stanisław Leśniewski. Grundzüge eines neuen Systems der Grundlagen der Mathematik. *Fundamenta Mathematicae*, 14:1–81, 1929.

Clarence Irving Lewis and Cooper Harold Langford. *Symbolic Logic*. New York, NY: Dover, second edition, 1959 [1932]. Republication of the first edition published by The Century Company in 1932.

David Lewis. Counterpart theory and quantified modal logic. *The Journal of Philosophy*, 65:113–126, 1968.

David Lewis. General semantics. *Synthese*, 22:18–67, 1970.

David Lewis. *Counterfactuals*. Oxford: Basil Blackwell, 1973.

David Lewis. Attitudes *De Dicto* and *De Se*. *The Philosophical Review*, 88:513–343, 1979.

David Lewis. Index, context, and content. In Stig Kanger and Sven Öhman, editors, *Philosophy and Grammar*, pages 79–100. Dordrecht: D. Reidel, 1980. Reprinted in his *Papers in Philosophical Logic*, pages 21–44. Cambridge: Cambridge University Press, 1998.

David Lewis. New work for a theory of universals. *Australasian Journal of Philosophy*, 61:343–377, 1983.

David Lewis. *On the Plurality of Worlds*. Oxford: Basil Blackwell, 1986.

Øystein Linnebo. Sets, properties, and unrestricted quantification. In Agustín Rayo and Gabriel Uzquiano, editors, *Absolute Generality*, pages 149–178. Oxford: Oxford University Press, 2006.

Øystein Linnebo. Plurals and modals. *Canadian Journal of Philosophy*, 46:654–676, 2016.

Øystein Linnebo and Agustín Rayo. Hierarchies ontological and ideological. *Mind*, 121:269–308, 2012.

John Mackay. Explaining the actuality operator away. *The Philosophical Quarterly*, 67: 709–729, 2017.

J. L. Mackie. *Truth, Probability and Paradox*. Oxford: Clarendon Press, 1973.

Antonella Mallozzi. What is absolute modality? *Inquiry*, forthcoming.

Benson Mates. Synonymity. In Leonard Linsky, editor, *Semantics and the Philosophy of Language*, pages 111–136. Urbana, IL: University of Illinois Press, 1952.

Christopher Menzel. The proper treatment of predication in fine-grained intensional logic. *Philosophical Perspectives*, 7:61–87, 1993.

Christopher Menzel. Possible worlds. In Edward N. Zalta, editor, *The Stanford Encyclopedia of Philosophy*. Metaphysics Research Lab, Stanford University, 2016.

Christopher Menzel and Edward N. Zalta. The fundamental theorem of world theory. *Journal of Philosophical Logic*, 43:333–363, 2014.

Richard Montague. *Formal Philosophy: Selected Papers of Richard Montague*. New Haven: Yale University Press, 1974. Edited and with an introduction by Richmond H. Thomason.

Reinhard Muskens. Intensional models for the theory of types. *The Journal of Symbolic Logic*, 72:98–118, 2007.

John Myhill. Problems arising in the formalization of intensional logic. *Logique et Analyse*, 1:78–83, 1958.

Daniel Nolan. The extent of metaphysical necessity. *Philosophical Perspectives*, 25: 313–339, 2011.

Daniel Nolan. Hyperintensional metaphysics. *Philosophical Studies*, 171:149–160, 2014.

Steven Orey. Model theory for the higher-order predicate calculus. *Transactions of the American Mathematical Society*, 92:72–84, 1959.

Barbara Hall Partee. Some structural analogies between tenses and pronouns in English. *The Journal of Philosophy*, 70:601–609, 1973.

Alvin Plantinga. *The Nature of Necessity*. Oxford: Clarendon Press, 1974.

Alvin Plantinga. Actualism and possible worlds. *Theoria*, 42:139–160, 1976.

Paul Portner. *Modality*. Oxford: Oxford University Press, 2009.

Lawrence Powers. COMMENTS on R. Stalnaker, "Propositions." In A. F. Mackay and D. D. Merrill, editors, *Issues in the Philosophy of Language*, pages 93–103. New Haven: Yale University Press, 1976.

Graham Priest. Metaphysical necessity: A skeptical perspective. *Synthese*, 198 (Suppl 8):S1873–S1885, 2021.

Arthur N. Prior. *Formal Logic*. Oxford: Clarendon Press, 1955.

Arthur N. Prior. Modality and quantification in S5. *The Journal of Symbolic Logic*, 21: 60–62, 1956.

Arthur N. Prior. *Time and Modality*. Oxford: Clarendon Press, 1957.

Arthur N. Prior. On a family of paradoxes. *Notre Dame Journal of Formal Logic*, 2: 16–32, 1961.

Arthur N. Prior. *Past, Present and Future*. Oxford: Clarendon Press, 1967.

Arthur N. Prior. Egocentric logic. *Noûs*, 2:191–207, 1968.

Ian Proops. What is Frege's 'concept horse problem'? In Peter Sullivan and Michael Potter, editors, *Wittgenstein's Tractatus: History and Interpretation*. Oxford: Oxford University Press, 2013.

Willard Van Orman Quine. Reference and modality. In *From a Logical Point of View: Nine Logico-Philosophical Essays*. Second Edition, Revised, pages 139–159. Cambridge, MA: Harvard University Press, 1980 [1953]. First edition published in 1953.

Willard Van Orman Quine. Propositional objects. In *Ontological Relativity and Other Essays*, pages 139–160. New York: Columbia University Press, 1969.

Agustín Rayo. On the open-endedness of logical space. *Philosophers' Imprint*, 20:1–21, 2020.

William N. Reinhardt. Satisfaction definitions and axioms of infinity in a theory of properties with necessity operator. In A. I. Arruda, R. Chuaqui, and N. C. A. da Costa, editors, *Mathematical Logic in Latin America*, pages 267–303. Amsterdam: North-Holland Publishing Company, 1980.

Gideon Rosen. The limits of contingency. In Fraser MacBride, editor, *Identity and Modality*, pages 13–39. Oxford: Oxford University Press, 2006.

Ian Rumfitt. Plural terms: Another variety of reference? In José Luis Bermúdez, editor, *Thought, Reference, and Experience: Themes from the Philosophy of Gareth Evans*, pages 84–123. Oxford: Oxford University Press, 2005.

Ian Rumfitt. Metaphysical dependence: Grounding and reduction. In Bob Hale and Aviv Hoffmann, editors, *Modality: Metaphysics, Logic, and Epistemology*, pages 34–64. Oxford: Oxford University Press, 2010.

Ian Rumfitt. *The Boundary Stones of Thought: An Essay in the Philosophy of Logic*. Oxford: Oxford University Press, 2015.

Bertrand Russell. *The Principles of Mathematics*. Cambridge: University Press, 1903.

Bertrand Russell. The philosophy of logical atomism. In R.C. Marsh, editor, *Logic and Knowledge*, pages 177–281. London: Allen & Unwin, 1956 [1918]. Transcript of lectures held in 1918, first published in *The Monist* in 1918–1919.

Nathan Salmon. *Frege's Puzzle*. Cambridge, MA: MIT Press, 1986.

Jonathan Schaffer. Confessions of a schmentencite: Towards an explicit semantics. *Inquiry*, 64:593–623, 2021.

Schiller Joe Scroggs. Extensions of the Lewis system S5. *The Journal of Symbolic Logic*, 16:112–120, 1951.

Stewart Shapiro. *Foundations without Foundationalism: A Case for Second-order Logic*. Oxford: Clarendon Press, 1991.

Theodore Sider. *Writing the Book of the World*. Oxford: Oxford University Press, 2011.

Theodore Sider. *The Tools of Metaphysics and the Metaphysics of Science*. Oxford: Oxford University Press, 2020.

Peter Simons. Stanisław Leśniewski. In Edward N. Zalta, editor, *The Stanford Encyclopedia of Philosophy*. Metaphysics Research Lab, Stanford University, 2020.

Jerzy Słupecki. St. Leśniewski's prototothetics. *Studia Logica*, 1:44–111, 1953.

Scott Soames. Direct reference, propositional attitudes, and semantic content. *Philosophical Topics*, 15:47–87, 1987.

Roy Sorensen. Vagueness. In Edward N. Zalta, editor, *The Stanford Encyclopedia of Philosophy*. Metaphysics Research Lab, Stanford University, 2022.

Robert Stalnaker. A theory of conditionals. In Nicholas Rescher, editor, *Studies in Logical Theory*, pages 98–112. Oxford: Blackwell, 1968.

Robert Stalnaker. Possible worlds. *Noûs*, 10:65–75, 1976.

Robert Stalnaker. *Inquiry*. Cambridge, MA: MIT Press, 1984.

Robert Stalnaker. *Context and Content: Essays on Intentionality in Speech and Thought*. Oxford: Clarendon Press, 1999.

Robert Stalnaker. *Ways a World Might Be*. Oxford: Clarendon Press, 2003.

Robert Stalnaker. *Mere Possibilities*. Princeton: Princeton University Press, 2012.

W. Starr. Counterfactuals. In Edward N. Zalta, editor, *The Stanford Encyclopedia of Philosophy*. Metaphysics Research Lab, Stanford University, 2019.

Roman Suszko. Identity connective and modality. *Studia Logica*, 27:7–41, 1971.

Roman Suszko. Abolition of the Fregean axiom. In Rohit Parikh, editor, *Logic Colloquium*, pages 169–239. Berlin: Springer-Verlag, 1975.

Alfred Tajtelbaum-Tarski. O wyrazie pierwotnym logistyki. *Przegląd Filozoficzny*, 26:68–89, 1923. Reprinted in English translation by J. H. Woodger as "On the Primitive Term of Logistic" in Alfred Tarski, *Logic, Semantics, Metamathematics*, pages 1–23. Oxford: Clarendon Press, 1956.

Leslie H. Tharp. The characterization of monadic logic. *The Journal of Symbolic Logic*, 38:481–488, 1973.

Robert Trueman. The concept *horse* with no name. *Philosophical Studies*, 172:1889–1906, 2015.

Gabriel Uzquiano. Plural quantification and modality. *Proceedings of the Aristotelian Society*, 111:219–250, 2011.

Bas C. van Fraassen. Facts and tautological entailments. *The Journal of Philosophy*, 66: 477–487, 1969.

Jean van Heijenoort. *From Frege to Gödel: A Source Book in Mathematical Logic, 1879–1931*. Cambridge, MA: Harvard University Press, 1967.

Peter van Inwagen. Modal epistemology. *Philosophical Studies*, 92:67–84, 1998.

Sean Walsh. Predicativity, the Russell-Myhill paradox, and Church's intensional logic. *Journal of Philosophical Logic*, 45:277–326, 2016.

Kai Frederick Wehmeier. In the mood. *Journal of Philosophical Logic*, 33:607–630, 2004.

Alfred North Whitehead and Bertrand Russell. *Principia Mathematica*, Volumes 1–3. Cambridge: Cambridge University Press, 1910–1913.

Timothy Williamson. The necessity and determinacy of distinctness. In Sabina Lovibond and S. G. Williams, editors, *Identity, Truth and Value: Essays for David Wiggins*, pages 1–17. Oxford: Blackwell, 1996.

Timothy Williamson. Everything. *Philosophical Perspectives*, 17:415–465, 2003.

Timothy Williamson. Conditionals and actuality. *Erkenntnis*, 70:135–150, 2009.

Timothy Williamson. Necessitism, contingentism, and plural quantification. *Mind*, 119:657–748, 2010.

Timothy Williamson. *Modal Logic as Metaphysics*. Oxford: Oxford University Press, 2013.

Timothy Williamson. Modal science. *Canadian Journal of Philosophy*, 46:453–492, 2016.

Ludwig Wittgenstein. Logisch-philosophische Abhandlung. In Wilhelm Oswald, editor, *Annalen der Naturphilosophie*, volume 14, pages 185–262. Leipzig: Unesma, 1921.

Seth Yalcin. Actually, *Actually*. *Analysis*, 75:185–191, 2015.

Juhani Yli-Vakkuri. Is meaning arbitrary? A philosopher looks at semiotics, 2018. URL https://ut.ee/en/content/inaugural-lecture-professor-juhani-yli-vakkuri-philosopher-looks-semiotics. Inaugural Lecture as Professor of the Philosophy of Language, University of Tartu, September 26th, 2018.

Juhani Yli-Vakkuri and John Hawthorne. Intensionalism and propositional attitudes. In Uriah Kriegel, editor, *Oxford Studies in Philosophy of Mind,* Volume 2, pages 114–174. Oxford: Oxford University Press, 2022.

List of Notation

Labeled Principles

Logics and Theories

Model-Theoretic Notation

Syntactic Mappings

Author Index

Subject Index

.